Essential Equations for the FE Exam Using the HP 33s

John A. Camara, PE

Professional Publications, Inc.
Belmont, CA

How to Locate and Report Errata for This Book

At Professional Publications, we do our best to bring you error-free books. But when errors do occur, we want to make sure you can view corrections and report any potential errors you find, so the errors cause as little confusion as possible.

A current list of known errata and other updates for this book is available on the PPI website at **www.ppi2pass.com/errata**. We update the errata page as often as necessary, so check in regularly. You will also find instructions for submitting suspected errata. We are grateful to every reader who takes the time to help us improve the quality of our books by pointing out an error.

ESSENTIAL EQUATIONS FOR THE FE EXAM USING THE HP 33s

Current printing of this edition: 1

Printing History

edition number	printing number	update
1	1	New book.

Printed in the United States of America

Professional Publications, Inc.
1250 Fifth Avenue, Belmont, CA 94002
(650) 593-9119
www.ppi2pass.com

ISBN-13: 978-1-59126-056-1
ISBN-10: 1-59126-056-6

The CIP data is pending.

TABLE OF CONTENTS

PROFESSIONAL PUBLICATIONS, INC.

ABOUT THE AUTHOR

John A. Camara holds a PE license in both electrical and nuclear engineering. He teaches a PE review course and the electrical engineering section of an FE/EIT review course, both for the University of Washington, Engineering Professional Program. He is a former associate technical fellow in electrical systems design, and currently works in engineering management of the Sea Launch program for the Boeing Commercial Space Company, a unit of The Boeing Company. A retired U.S. Navy lieutenant commander, he served as a nuclear-trained electrical engineer and submarine officer. Mr. Camara received his bachelor of science degree in electrical and computer engineering/materials science and engineering from the University of California at Davis and his master of science degree in space systems from the Florida Institute of Technology.

PREFACE

Doing well on the FE/EIT examination isn't just about what you know. You have to have the knowledge, of course, but just as important are accuracy and speed. If you can solve problems quickly, and make fewer errors in the process, then you'll score higher on the exam.

The hp 33s calculator, which the NCEES allows you to use on the exam, makes it possible for you to store equations in memory prior to the exam, and then use them during it to increase your speed and accuracy. That's the reason for this book.

I've gone through the NCEES *Fundamentals of Engineering Supplied-Reference Handbook*—over 200 pages—and picked the 38 equations I think are most useful to have stored in the hp 33s during the exam. However, if you're taking the chemical PM session, you will have a different focus than someone taking the electrical session. To help you pick out the equations you need most, I've labeled each one with the session or sessions in which you are most likely to encounter it.

Then I worked out the sequence of keystrokes for entering each equation into the hp 33s memory, so you can quickly store the ones you want. Additionally, I've picked out practice problems to show how to use stored equations to solve problems with greater speed and accuracy.

Using these stored equations, and this book, will enable you to calculate results faster. No matter how complex the equation is, all you need to do is enter the values of the known variables. You'll be more accurate because the calculator does the rest of the work.

I hope you enjoy the process of learning as you pursue your success on the exam.

John A. Camara, PE

ACKNOWLEDGMENTS

I cannot thank enough the fine people at Professional Publications, both past and present, for their help and encouragement from my first book to this one. The staff at PPI contributed many great ideas at the start of this effort, and I hope the end product meets their vision.

Sarah Hubbard provided her always insightful input and made me a believer in the project from day one. Marjorie Roueche set the stage and defined the scope of the work with precision and encouragement, for which I'm grateful. Scott Marley had the unenviable task of finding the errors and creating a smooth product from the rough, and I thank him for it. Thanks as well to Cathy Schrott, who designed the layout, Miriam Hanes, who executed it, and Tom Bergstrom, who handled the illustrations.

I also wish to acknowledge the contributions of the authors whose books appear in the list of references, as it is their problems I have used and adapted throughout this book. I also thank Hewlett-Packard for allowing the use of many figures from the hp 33s User's Manual and for supporting this project.

Finally, my wife Becky is to be thanked for all that she does for me and for the sunshine that she brings into my life. Early in our relationship, I remarked to a relative that Becky "makes my soul smile." It is still true today.

John A. Camara, PE

INTRODUCTION

HOW TO USE THIS BOOK

The hp 33s is a scientific calculator approved by the NCEES for use during the FE/EIT examination. This book will help you save time on the exam by giving you practice with the calculator's Equation mode and SOLVE function, concentrating on the types of equations and problems you are likely to encounter.

Faced with a problem concerning half-life, for example, you might need a couple of minutes to calculate the value of

$$N = N_0 e^{-0.0693t/\tau}$$

And you'd probably need even longer if the variable you needed to calculate was τ and not N. But if you have this equation already stored in the hp 33s, you only need to enter the values of N, N_0, and t, and you have the value of τ in a moment.

This book will help you decide which equations to store. When you turn to page 1, you'll see a list of 38 of the most common equations on the FE/EIT exam. Next to each equation is the page number on which it appears in the NCEES *Fundamentals of Engineering Supplied-Reference Handbook*, 7th Edition. Also shown are the subject disciplines each equation is most likely to appear under during the examination. This will help you determine which equations to store.

The rest of the book is in four parts. Part A, the largest, shows you how to store each of these equations in the calculator. The steps for entering Equation 1 are explained in particular detail, so it's a good place to start if you're unfamiliar with the process.

Part B shows how to use these stored equations to save time in solving problems. Part C demonstrates the use of some preprogrammed functions of the hp 33s that can be useful on the exam. Finally, Part D is for those who prefer Algebraic mode to Reverse Polish Notation mode, and it covers the most important differences between these two modes.

The hp 33s has sufficient memory to store all the equations in this book. However, the more equations you store, the longer it will take you to find a particular equation during the examination. You may find it best to store only those equations most expected to be useful to you. Taking a practice exam using the hp 33s can help you gauge the number of equations you can comfortably handle on your list.

In Equation mode, equations are entered in algebraic notation. The usual operator precedence is followed: first operators in parentheses, then exponents, the unary minus (changing the sign of a number), multiplication and division, and finally addition and subtraction. Multiplication by adjacent parentheses is not recognized; the multiplication symbol must be used.

After you have entered an equation, you can display a unique checksum and the length of the equation by pressing and holding the SHOW key. If the numbers shown on the calculator match the numbers in this book, the equation has been entered properly.

The display of the hp 33s has two lines. Whenever only one line is given in this book, it is the lower that is shown.

The hp 33s cannot represent variables in lower case, with subscripts, or containing multiple letters. Normally in this book a letter will be used similar to that in the NCEES equation. A Greek letter is represented either by its English transliteration (as B for β, beta), or by an English letter that resembles it reasonably closely (as P for ρ, rho). Where two variables use the same letter with subscripts 1 and 2, the preceding letter is used for the former. Thus, T is used for T_2 and S is used for T_1.

You may use other variables if you prefer. However, if you change a variable, or a sequence of operations, the checksum will differ from that shown here.

When an equation being solved has multiple roots, the calculator may return zero, find no root, or show the correct answer only after a considerable time. To avoid this, store a reasonable value in the variable of interest before solving the equation. Doing so bounds the problem. On the exam, you can determine a reasonable value from the possible answers given for a problem. See problem 3 in Part B for more information or Chapter 7 and Appendix D in the User's Manual.

When an equation being entered becomes longer than can be displayed on the hp 33s, a small arrow appears on the left side of the display (see Annunciators later in this introduction). By scrolling with the large silver key just below the display, one can view any portion of the equation desired.

GETTING STARTED WITH THE hp 33s

If you are not familiar with the hp 33s and your time is limited, here is a brief overview, focusing on what you will need to know to use the calculator during the exam. Other specific functions will be explained as they come up in the book. If you want to learn more, Appendix G in the User's Manual defines and explains the use of each and every key and function of the hp 33s.

The topography of the hp 33s is shown here. The calculator is turned on by pressing ⌨ **C** in the lower left corner. The calculator is turned off by pressing the purple 🔁 key (located just above the **C** key), followed by **C**.

Most keys are associated with three functions. The main function is printed in white on the face of the key. These functions include the number keys, the operators ⊞, ⊟, ⊠, and ⊡, and so on. To use a key's main function, press the key without pressing a shift key first.

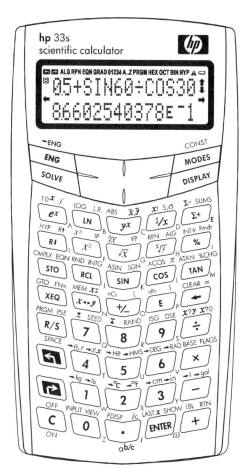

Shifted Keys

Above each key are printed two more functions, one in green and one in purple. These functions are accessed by first pressing the green or purple shift key (⬛ or ⬛), and then pressing the key for the function. Pressing a shift key two times in a row cancels the shift.

Some green and purple functions are grouped by field. Statistical functions are in the top row, trigonometric functions are in the third row, and probability functions are in the fourth.

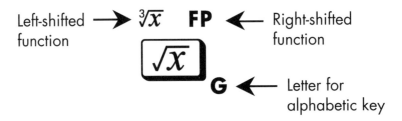

Alpha Keys

Some keys also have a letter of the alphabet printed to the lower right. Use these "alpha keys" whenever the hp 33s expects a variable (or other alphabetic label). Pressing `RCL` `K`, for example, puts the current value of variable K into the display; whatever key is pressed next after the `RCL` key is treated by the HP 33s as a variable. When the alpha keys are active, the small **A..Z** annunciator appears at the top of the display.

Silver Keys

The large silver key is used to move through various lists or answers. This key can be thought of as having arrows at its indentations. Press on the top edge to scroll up, on the right edge to scroll to the right, and so on.

When there is information off the display that can be brought into view with the large silver key, small arrows will appear on the screen.

PROFESSIONAL PUBLICATIONS, INC.

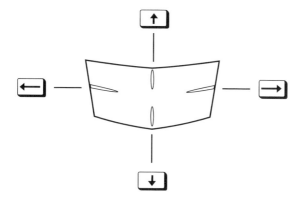

COPYRIGHT 2003 HEWLETT-PACKARD DEVELOPMENT COMPANY, L.P. REPRODUCED WITH PERMISSION.

To the sides of the large silver key are four smaller silver keys. The pressure points on the keys are shown.

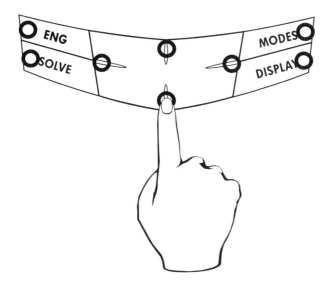

COPYRIGHT 2003 HEWLETT-PACKARD DEVELOPMENT COMPANY, L.P. REPRODUCED WITH PERMISSION.

Annunciators

Small symbols called annunciators will appear on the display at times, usually to give information about the mode or status that the hp 33s is in. All the annunciators are shown.

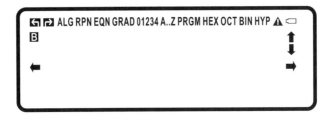

Correcting Errors

To correct errors, backspace with the ⟵ key, or clear the number currently displayed with the C key. These keys and the CLEAR key are explained in more detail on pages 1-5 and 1-6 of the User's Manual.

RPN and ALG Modes

The calculator is normally in the RPN mode. If it is not, it can be set by pressing ◂ RPN. The ALG mode may be selected by pressing ▸ ALG.

EQUATION MODE VERSUS PROGRAMMING

These instructions show how to use the calculator's Equation mode, not its real programming capabilities. Programming per se is covered in the User's Manual, chapters 12 to 17.

Programming allows you to use word prompts, and lets you retrieve an equation with a letter name instead of by scrolling through a list. But these are small gains, and programming is time-consuming and tedious. The goal of this text is to prepare you for the exam by the most expeditious path, and that path is Equation mode.

REFERENCES

Problems and solutions in this book have been selected and adapted from the following books. All are from Professional Publications, Inc.

Angus, Robert B., et al. *Electrical Discipline-Specific Review for the FE/EIT Exam.*

Camara, John A. *Electrical Engineering Reference Manual for the Electrical and Computer PE Exam.*

Kim, Robert H., with Michael R. Lindeburg. *Civil Discipline-Specific Review for the FE/EIT Exam.*

Lindeburg, Michael R. *FE Review Manual: Rapid Preparation for the General Fundamentals of Engineering Exam.*

Lindeburg, Michael R. *1001 Solved Engineering Fundamentals Problems.*

Lopina, Stephanie T., with Michael R. Lindeburg. *Chemical Discipline-Specific Review for the FE/EIT Exam.*

Noble, James S., et al. *Industrial Discipline-Specific Review for the FE/EIT Exam.*

Saad, Michel, et al. *Mechanical Discipline-Specific Review for the FE/EIT Exam.*

LIST OF EQUATIONS

This book contains keystroke-by-keystroke instructions for entering 38 equations into the hp 33s calculator. These equations are listed here, along with some information that will help you determine which ones will be most useful to you.

Following the name of each equation is the page number where the equation is found in the NCEES *Fundamentals of Engineering Supplied-Reference Handbook*, 7th Edition.

Shown to the right of each equation is a list of the exam sessions in which it is most likely to appear. (All examinees take the same morning session and choose from among seven possible afternoon sessions: chemical, civil, electrical, environmental, general, industrial, and mechanical.) At best, this list cannot be anything more than an educated guess, and what was true in previous exams may or may not be true of the one you take. Use your own judgment.

Engineering Probability and Statistics

1. Binomial Distribution (Handbook p. 16)   AM, Gen, Ind

$$P(x) = \left(\frac{n!}{x!\,(n-x)!} \right) p^x q^{n-x}$$

Statics

2. Resolution of a Force (Handbook p. 24)   AM, Chem, Mech

$$R = \sqrt{x^2 + y^2 + z^2}$$

Dynamics

3. Constant Acceleration (Handbook p. 30)   AM, Gen, Mech

$$s = s_0 + v_0 t + \frac{a_0 t^2}{2}$$

Mechanics of Materials

4. Mohr's Circle (Handbook p. 39)   AM, Mech

$$R = \sqrt{\left(\frac{\sigma_x - \sigma_y}{2}\right)^2 + \tau_{xy}^2}$$

5. Torsional Strain (Handbook p. 41)   AM, Gen, Mech

$$\phi = \frac{TL}{GJ}$$

6. Beam Deflection (Handbook p. 43)   AM, Gen, Mech

$$\delta = \frac{Px^2}{6EI}(-x + 3a) \quad \text{(for } x \le a)$$

Fluid Mechanics

7. Continuity Equation (Handbook p. 45)   AM, Civ, Env, Mech

$$\rho_2 A_2 v_2 = \rho_1 A_1 v_1$$

8. Field Equation/Incompressible Flow   AM, Civ, Env, Mech
    (Bernoulli Equation) (Handbook pp. 46, 47)

$$\frac{p_2}{\gamma} + z_2 + \frac{v_2^2}{2g} + h_f = \frac{p_1}{\gamma} + z_1 + \frac{v_1^2}{2g}$$

9. Reynolds Number (Handbook p. 47)   AM, Chem, Env, Gen,
                                                       Mech
$$Re = \frac{vD\rho}{\mu}$$

10. Reynolds Number (Handbook p. 47)   AM, Chem, Env, Gen,
                                                       Mech
$$Re = \frac{VD}{\upsilon}$$

11. Darcy-Weisbach Equation (Handbook p. 47)   Civ, Env, Gen, Mech

$$h_f = f\frac{Lv^2}{D2g}$$

12. Venturi Meters (Handbook p. 50)   Civ, Env, Gen, Mech

$$Q = \frac{C_v A_2}{\left(\sqrt{1 - \left(\frac{A_2}{A_1}\right)^2}\right)}\left(\sqrt{2g\left(\frac{p_1}{\gamma} + z_1 - \frac{p_2}{\gamma} - z_2\right)}\right)$$

Thermodynamics

13. Ideal Gases (Handbook p. 56)

$$pv = RT$$

AM, Chem, Env, Gen, Mech

14. Ideal Gases (Handbook p. 56)

$$pV = mRT$$

AM, Chem, Env, Gen, Mech

15. Ideal Gases (Handbook p. 56)

$$\frac{p_1 v_1}{T_1} = \frac{p_2 v_2}{T_2}$$

AM, Chem, Env, Gen, Mech

16. Constant Entropy Processes (Handbook p. 56)

$$p_1 v_1^k = p_2 v_2^k$$

AM, Gen, Mech

17. Constant Entropy Processes (Handbook p. 56)

$$T_1 p_1^{\frac{1-k}{k}} = T_2 p_2^{\frac{1-k}{k}}$$

AM, Gen, Mech

18. Carnot Cycle Efficiency (Handbook p. 58)

$$\eta_c = \frac{(T_H - T_L)}{T_H}$$

AM, Chem, Elec, Gen, Mech

Heat Transfer

19. Conduction Through a Plane Wall (Handbook p. 67)

$$\dot{Q} = \frac{-kA(T_2 - T_1)}{L}$$

AM, Gen, Mech

20. Conduction Through a Cylindrical Wall (Handbook p. 67)

$$\dot{Q} = \frac{2\pi kL(T_1 - T_2)}{\ln\left(\dfrac{r_2}{r_1}\right)}$$

Gen, Mech

21. Radiation Emitted by a Body (Handbook p. 68)

$$\dot{Q} = \varepsilon \sigma A T^4$$

AM, Gen, Mech

22. Black Body Radiation Energy Exchange   Gen, Mech
    (Handbook p. 68)

$$\dot{Q}_{12} = A_1 F_{12} \sigma \left(T_1^4 - T_2^4 \right)$$

23. Body Immersed in Flowing Fluid   Chem, Mech
    (Handbook p. 69)

$$\mathrm{Nu} = c\mathrm{Re}^n \mathrm{Pr}^{1/3}$$

Transport Phenomena

24. Nusselt Number (Handbook p. 73)   Chem, Mech

$$\mathrm{Nu} = \frac{hD}{k}$$

25. Prandtl Number (Handbook p. 73)   Chem, Mech

$$\mathrm{Pr} = \frac{c_P \mu}{k}$$

Chemistry

26. Acids and Bases (Handbook p. 78)   AM, Chem, Env

$$\mathrm{pH} = \log_{10} \left(\frac{1}{[\mathrm{H}^+]} \right)$$

27. Faraday's Law (Handbook p. 78)   AM, Chem, Elec, Gen

$$m_{\mathrm{grams}} = \frac{It(\mathrm{MW})}{(96{,}485)\,(\text{change in oxidation state})}$$

Materials Science/Structure of Matter

28. Diffusion Coefficient (Handbook p. 82)   AM, Chem, Gen, Mech

$$D = D_0 e^{-Q/RT}$$

29. Half-Life (Handbook p. 85)   AM, Chem, Env, Gen

$$N = N_0 e^{-0.0693t/\tau}$$

Engineering Economics

30. Single Payment Compound Amount   AM, Chem, Civ, Elec,
    (Handbook p. 92)   Gen, Ind, Mech

$$(F/P, i\%, n) = (1 + i)^n$$

31. Capital Recovery (Handbook p. 92)

$$(A/P, i\%, n) = \frac{i(1+i)^n}{((1+i)^n - 1)}$$

AM, Chem, Civ, Elec, Gen, Ind, Mech

32. Non-annual Compounding (Handbook p. 92)

$$i_e = \left(1 + \frac{r}{m}\right)^m - 1$$

AM, Chem, Civ, Elec, Gen, Ind, Mech

Civil Engineering

33. Manning's Equation (Handbook p. 137)

$$Q = \frac{K}{n} AR^{2/3} S^{1/2}$$

Civ

Electrical and Computer Engineering

34. Electrostatic Force (Handbook p. 167)

$$F = \frac{Q_1 Q_2}{4\pi \varepsilon r^2}$$

AM, Elec, Gen

35. Electric Field Intensity (Handbook p. 167)

$$E = \frac{Q_1}{4\pi \varepsilon r^2}$$

AM, Elec, Gen

36. Two Resistors in Parallel, or Two Capacitors in Series (Handbook p. 168)

$$R_P = \frac{R_1 R_2}{(R_1 + R_2)}$$

$$C_S = \frac{1}{\dfrac{1}{C_1} + \dfrac{1}{C_2}} = \frac{C_1 C_2}{C_1 + C_2}$$

AM, Elec, Gen

37. Energy Stored in an Inductor (Handbook p. 168)

$$U = \frac{L i_L^2}{2}$$

AM, Elec, Gen

38. Resonance Frequency (Handbook p. 170)

$$\omega_0 = \frac{1}{\sqrt{LC}}$$

AM, Elec, Gen

STORING EQUATIONS

ENGINEERING PROBABILITY AND STATISTICS

Equation 1: Binomial Distribution (Handbook p. 16)

$$P\left(x\right) = \left(\frac{n!}{x!\left(n-x\right)!}\right) p^x q^{n-x}$$

Entered as

B=(χ ! (N) ÷ (χ ! (X) × χ ! (N–X))) ×P^X×Q^(N–X)

Sessions AM, PM

Disciplines General, Industrial

Variables The hp 33s can only recognize single uppercase Roman letters as variables. $P(x)$ cannot, then, be used. Also, $P(x)$ cannot simply be replaced with P, because P and p must be distinguished. In these instructions, $P(x)$ is replaced with B, for binomial. If preferred, different letters can be assigned to the variables, but will lead to a different checksum at the end.

Factorials The factorial key is entered before the variable, not after. A factorial is displayed on the calculator as the symbol χ! followed by the argument in parentheses. χ here is not a variable but part of the symbol. Thus, the expression $n!$ appears on the hp 33s display as χ!(N).

Keys	Display	Description
⏩	0.0000	Selects right-shift (purple) functions. Display will read 0.0000 if calculator is set to four decimal places and contains a value of zero. Continue in any case.

EQN	EQN LIST TOP	Selects Equation mode. **EQN** annunciator appear. If this is not the first equation entered, other equations may be displayed. Continue regardless.
RCL	■	Starting new equation causes rectangular cursor to appear. Pressing RCL causes **A..Z** annunciator to appear. A variable can now be entered.
B	B■	Enters variable B, representing $P(x)$.
↰ =	B=■	
↰ (	B=(■	
↰ $x!$	B=(χ!(■	Opens factorial function.
RCL N	B=(χ!(N■	Enters variable N.
↰)	B=(χ!(N) ■	Closes factorial function.
÷	B=(χ!(N)÷■	
↰ (	B=(χ!(N)÷(■	Multiplication and division have the same "operator precedence." Therefore, parentheses, which have the highest precedence, are used to ensure the equation is evaluated correctly.
↰ $x!$	B=(χ!(N)÷(χ!(■	Opens factorial function.
RCL X	=(χ!(N)÷(χ!(X■	Enters variable X. Equation is now longer than the hp 33s display. For instructions on scrolling left and right, see How to Use This Book.
↰)	(χ!(N)÷(χ!(X) ■	Closes factorial function.

PROFESSIONAL PUBLICATIONS, INC.

$\boxed{\times}$	$\chi!(N)\div(\chi!(X)\times\blacksquare$	The hp 33s does not allow implied multiplication using adjacent parentheses. The multiplication operator must be used.
$\boxed{\leftarrow}$ $\boxed{x!}$	$N)\div(\chi!(X)\times\chi!(\blacksquare$	Opens factorial function.
$\boxed{\text{RCL}}$ $\boxed{N}$	$)\div(\chi!(X)\times\chi!(N\blacksquare$	Enters variable N.
$\boxed{-}$	$\div(\chi!(X)\times\chi!(N-\blacksquare$	
$\boxed{\text{RCL}}$ $\boxed{X}$	$(\chi!(X)\times\chi!(N-X\blacksquare$	Enters variable X.
$\boxed{\rightarrow}$ $\boxed{)}$	$\chi!(X)\times\chi!(N-X)\blacksquare$	Closes factorial function. The expression $(n-x)!$ appears as $\chi!(N-X)$.
$\boxed{\rightarrow}$ $\boxed{)}$	$!(X)\times\chi!(N-X))\blacksquare$	
$\boxed{\rightarrow}$ $\boxed{)}$	$(X)\times\chi!(N-X)))\blacksquare$	
$\boxed{\times}$	$X)\times\chi!(N-X)))\times\blacksquare$	
$\boxed{\text{RCL}}$ $\boxed{P}$	$)\times\chi!(N-X)))\times P\blacksquare$	
$\boxed{y^x}$	$\times\chi!(N-X)))\times P\wedge\blacksquare$	The operator ^ indicates an exponent. The expression p^x appears as $P\wedge X$.
$\boxed{\text{RCL}}$ $\boxed{X}$	$\chi!(N-X)))\times P\wedge X\blacksquare$	Exponents are evaluated before multiplication and division. Therefore, parentheses are not required to ensure correct precedence.
$\boxed{\times}$	$!(N-X)))\times P\wedge X\times\blacksquare$	
$\boxed{\text{RCL}}$ $\boxed{Q}$	$(N-X)))\times P\wedge X\times Q\blacksquare$	
$\boxed{y^x}$	$N-X)))\times P\wedge X\times Q\wedge\blacksquare$	
$\boxed{\rightarrow}$ $\boxed{(}$	$-X)))\times P\wedge X\times Q\wedge(\blacksquare$	

RCL N	X)))×P^X×Q^(N∎	
−	)))×P^X×Q^(N−∎	
RCL X	))×P^X×Q^(N−X∎	
⮕ ⏺	)×P^X×Q^(N−X)∎	
ENTER	B=(χ!(N)÷(χ!(X	Ends equation. For instructions on scrolling left and right, see How to Use This Book.
⮕ SHOW	CK=E389 LN=37	Press and hold SHOW to display checksum and length of equation. If numbers displayed match these, equation is correctly entered.

STATICS

Equation 2: Resolution of a Force (Handbook p. 24)

$$R = \sqrt{x^2 + y^2 + z^2}$$

Entered as

R=SQRT(SQ(X)+SQ(Y)+SQ(Z))

Sessions AM, PM

Disciplines Chemical, Mechanical

Variables The hp 33s cannot distinguish between uppercase and lowercase. All variables appear as capital letters.

Keys	Display	Description
⮕ EQN RCL R ⮕ =	R=∎	Selects Equation mode and starts new equation.
√x̄	R=SQRT(∎	Opens square root function. Functions can be nested within other functions.

x^2 RCL X	R=SQRT(SQ(X■	Opens square function and enters X.
⮐)	R=SQRT(SQ(X)■	Closes square function. The expression x^2 appears as $SQ(X)$.
+	R=SQRT(SQ(X)+■	
x^2 RCL Y ⮐)	T(SQ(X)+SQ(Y)■	Enters Y^2. Equation is now longer than the hp 33s display. For instructions on scrolling left and right, see How to Use This Book.
+	(SQ(X)+SQ(Y)+■	
x^2 RCL Z ⮐)	)+SQ(Y)+SQ(Z)■	Enters Z^2.
⮐)	+SQ(Y)+SQ(Z))■	Closes square root function.
ENTER	R=SQRT(SQ(X)+S■	Ends equation.
⮐ SHOW	CK=C973 LN=25	Press and hold SHOW to display checksum and length of equation. If numbers displayed match these, equation is correctly entered.

DYNAMICS

Equation 3: Constant Acceleration (Handbook p. 30)

$$s = s_0 + v_0 t + \frac{a_0 t^2}{2}$$

Entered as

S=D+V×T+A×SQ(T)÷2

Sessions AM, PM

Disciplines General, Mechanical

Variables The hp 33s can only recognize single letters as variables, so subscripts cannot be used to distinguish s and s_0. Replace s_0 with D, for initial distance.

If preferred, different letters can be assigned to the variables, but will lead to a different checksum at the end.

Keys	Display	Description
[◄] [EQN] [RCL] [S] [◄] [=]	S=∎	Selects Equation mode and starts new equation. *S* represents the straight line distance (*s*).
[RCL] [D]	S=D∎	*D* represents initial distance (*s₀*).
[+] [RCL] [V]	S=D+V∎	*V* represents initial velocity (*v₀*).
[×] [RCL] [T]	S=D+V×T∎	*T* represents time (*t*).
[+] [RCL] [A]	S=D+V×T+A∎	*A* represents acceleration (*a*).
[×] [x^2]	S=D+V×T+A×SQ(∎	Opens square function.
[RCL] [T] [◄] [)]	D+V×T+A×SQ(T) ∎	Enters *T* and closes square function. Equation is now longer than the hp 33s display. For instructions on scrolling left and right, see How to Use This Book.
[÷] [2]	×T+A×SQ(T)÷2_	Cursor changes for entering digits.
[ENTER]	S=D+V×T+A×SQ(T∎	Ends equation.
[◄] [SHOW]	CK=FED0 LN=17	Press and hold SHOW to display checksum and length of equation. If numbers displayed match these, equation is correctly entered.

MECHANICS OF MATERIALS

Equation 4: Mohr's Circle (Handbook p. 39)

$$R = \sqrt{\left(\frac{\sigma_x - \sigma_y}{2}\right)^2 + \tau_{xy}^2}$$

Entered as

R=SQRT(SQ((X−Y)÷2)+SQ(T))

Sessions AM, PM

Disciplines Mechanical

Variables The hp 33s can only recognize Roman letters as variables. X and Y represent σ_x and σ_y, respectively, and T represents τ. If preferred, different letters can be assigned to the variables, but will lead to a different checksum at the end.

Keys	Display	Description
◤ EQN RCL R ◤ =	R=■	Selects Equation mode and starts new equation. *R* represents the radius of Mohr's circle.
√x̄	R=SQRT(■	Opens square root function.
x²	R=SQRT(SQ(■	Opens square function. Functions can be nested within other functions.
◤ () RCL X	R=SQRT(SQ((X■	*X* stands for σ_x.
− RCL Y ◤)	SQRT(SQ((X−Y) ■	Equation is now longer than the hp 33s display. For instructions on scrolling left and right, see How to Use This Book.
÷	QRT(SQ((X−Y)÷ ■	
2	T(SQ((X−Y)÷ 2_	Cursor changes for entering digits.
◤)	T(SQ((X−Y)÷2) ■	Closes square function.
+ x² RCL T	(X−Y)÷2)+SQ(T■	Opens square function. *T* represents shear stress (τ).
◤) ◤)	(Y)÷2)+SQ(T)) ■	Closes square and square root functions.
ENTER	R=SQRT(SQ((X−Y	Ends equation.

| 🠒 SHOW | CK=AF9F
LN=25 | Press and hold SHOW to display checksum and length of equation. If numbers displayed match these, equation is correctly entered. |

Equation 5: Torsional Strain (Handbook p. 41)

$$\phi = \frac{TL}{GJ}$$

Entered as

P=T×L÷G÷J

Sessions AM, PM

Disciplines General, Mechanical

Variables The hp 33s can only recognize Roman letters as variables. P represents the Greek letter ϕ (phi). If preferred, different letters can be assigned to the variables, but will lead to a different checksum at the end.

Keys	Display	Description
🠒 EQN RCL P 🠒 =	P=∎	Selects Equation mode and starts new equation. P represents angle of twist (ϕ).
RCL T	P=T∎	T represents torque.
× RCL L	P=T×L∎	L represents length.
÷ RCL G	P=T×L÷G∎	G represents shear modulus.
÷ RCL J	P=T×L÷G÷J∎	J represents polar moment of inertia.
ENTER	P=T×L÷G÷J	Ends equation.
🠒 SHOW	CK=007B LN=9	Press and hold SHOW to display checksum and length of equation. If numbers displayed match these, equation is correctly entered.

14

Equation 6: Beam Deflection (Handbook p. 43)

$$\delta = \frac{Px^2}{6EI}(-x + 3a) \quad \text{(for } x \leq a)$$

δ is positive downward.

Entered as

D=P×SQ(X)×(−X+3×A)÷6÷E÷I

Sessions AM, PM

Disciplines General, Mechanical

Variables The hp 33s can only recognize single Roman letters as variables. The Greek letter δ (delta) is replaced with D, for deflection. If preferred, different letters can be assigned to the variables, but will lead to a different checksum at the end.

Keys	Display	Description
[⇄] [EQN] [RCL] [D] [⇄] [=]	D=∎	Selects Equation mode and starts new equation. D represents deflection (δ).
[RCL] [P]	D=P∎	P represents load.
[×] [x²] [RCL] [X] [⇄] [)]	D=P×SQ(X)∎	Enters X^2.
[×] [⇄] [(] [+/−]	D=P×SQ(X)×(−∎	Changes sign of next number.
[RCL] [X] [+]	=P×SQ(X)×(−X+∎	Equation is now longer than the hp 33s display. For instructions on scrolling left and right, see How to Use This Book.
[3]	×SQ(X)×(−X+ 3_	Cursor changes for entering digits.
[×] [RCL] [A]	SQ(X)×(−X+3×A∎	Enters variable A.
[⇄] [)] [÷]	(X)×(−X+3×A)÷∎	
[6]	)×(−X+3×A)÷ 6_	Cursor changes for entering digits.
[÷] [RCL] [E]	×(−X+3×A)÷6÷E∎	E represents modulus of elasticity.

÷ RCL I	−X+3×A)÷6÷E÷I∎	I represents moment of inertia.
ENTER	D=P×SQ(X)×(−X+	Ends equation.
⤶ SHOW	CK=D9D5 LN=24	Press and hold SHOW to display checksum and length of equation. If numbers displayed match these, equation is correctly entered.

FLUID MECHANICS

Equation 7: Continuity Equation (Handbook p. 45)

$$\rho_2 A_2 v_2 = \rho_1 A_1 v_1$$

Entered as

P×A×V=O×Z×U

Sessions AM, PM

Disciplines Civil, Environmental, Mechanical

Variables The hp 33s can only recognize Roman letters as variables. P represents the Greek letter ρ (rho), because of the resemblance. To distinguish variables with subscript 1 from those with subscript 2, P, A, and V represent ρ_2, A_2, and v_2, respectively. For ρ_1, A_1, and v_1, each letter is shifted back one place in the alphabet and O, Z, and U, respectively, are used. If preferred, different letters can be assigned to the variables, but will lead to a different checksum at the end.

Solving for mass flow rate This equation can also be used to calculate mass flow rate, from the equation $\dot{m} = \rho A v$. Solve for O, and set Z and U each equal to 1. Enter values for P, A, and V as usual.

Keys	Display	Description
⤶ EQN RCL P	P∎	Selects Equation mode and starts new equation. *P* represents density (ρ_2).
× RCL A	P×A∎	*A* represents cross-sectional area (A_2).

☒ RCL V	P×A×V▮	*V* represents velocity (v₂).
⬌ = RCL O	P×A×V=O▮	*O* stands for ρ_2, as explained above.
☒ RCL Z	P×A×V=O×Z ▮	*Z* stands for A_1.
☒ RCL U	P×A×V=O×Z×U▮	*U* stands for v_1.
ENTER	P×A×V=O×Z×U	Ends equation.
⬌ SHOW	CK=6D92 LN=11	Press and hold SHOW to display checksum and length of equation. If numbers displayed match these, equation is correctly entered.

Equation 8: Field Equation/Energy Equation for Incompressible Flow (Bernoulli Equation) (Handbook p. 46, 47)

$$\frac{p_2}{\gamma} + z_2 + \frac{v_2^2}{2g} + h_f = \frac{p_1}{\gamma} + z_1 + \frac{v_1^2}{2g}$$

Entered as

P÷G+Z+SQ(V)÷19.62+H=O÷G+Y+SQ(U)÷19.62

Sessions AM, PM

Disciplines General, Civil, Environmental, Mechanical

Using this equation This equation serves a dual purpose: It can be used to solve problems with or without head loss. If there are no friction losses, enter zero as the value for *H*. That will make this equation equivalent to the *field equation* on p. 46 of the Handbook. For problems in *real fluid flow* or *steady, incompressible flow* in conduits and pipes, *H* stands for h_f.

Variables The hp 33s can only recognize single Roman letters as variables. The Greek letter γ (gamma) is replaced with *G*. (*g*, a constant, is replaced with its numerical value, so there is no conflict.) To distinguish variables with subscript 1 from those with subscript 2, *P*, *V*, and *Z* are used for p_2, v_2, and z_2, respectively. For p_1, v_1, and z_1, each letter is shifted back one place in the alphabet and *O*, *U*, and *Y*, respectively, are used. If preferred, different letters can be assigned to the variables, but will lead to a different checksum at the end.

Keys	Display	Description
⬅ EQN RCL P	P ∎	Selects Equation mode and starts new equation. P represents pressure (p_2).
÷ RCL G	P÷G ∎	G represents specific weight (γ).
+ RCL Z	P÷G+Z ∎	Z represents elevation (z_2).
+ x^2	P÷G+Z+SQ(∎	Opens square function.
RCL V	P÷G+Z+SQ(V ∎	V represents velocity (v_2).
⬅) ÷	P÷G+Z+SQ(V)÷ ∎	Closes square function.
1 9 · 6 2	+SQ(V)÷ 19.62_	19.62 is the value of 2g. Cursor changes for entering digits. Equation is now longer than the hp 33s display. For instructions on scrolling left and right, see How to Use This Book.
+ RCL H	SQ(V)÷19.62+H ∎	H represents head loss (h_f).
⬅ = RCL O	(V)÷19.62+H=O ∎	O stands for p_1, as explained above.
÷ RCL G	)÷19.62+H=O÷G ∎	
+ RCL Y	19.62+H=O÷G+Y ∎	Y stands for z_1.
+ x^2 RCL U ⬅) ÷	=O(G+Y+SQ(U)÷ ∎	U stands for v_1.
1 9 · 6 2	+SQ(U)÷ 19.62_	
ENTER	P÷G+Z+SQ(V)÷19	Ends equation.

🔁 SHOW	CK=5840 LN=37	Press and hold SHOW to display checksum and length of equation. If numbers displayed match these, equation is correctly entered.

Equation 9: Reynolds Number (Handbook p. 47)

$$\text{Re} = \frac{\text{v}D\rho}{\mu}$$

Entered as

R=V×D×P÷M

Sessions PM

Disciplines General, Chemical, Environment, Mechanical

Variables The hp 33s can only recognize single Roman letters as variables. R is used for Re; P for the Greek letter ρ (rho), because of the physical similarity; and M for μ, the Greek letter mu. If preferred, different letters can be assigned to the variables, but will lead to a different checksum at the end.

Keys	Display	Description
🔁 EQN RCL R 🔁 =	R= ∎	Selects Equation mode and starts new equation. R represents the Reynolds number (Re).
RCL V	R=V∎	V represents velocity (v).
× RCL D	R=V×D∎	D represents diameter.
× RCL P	R=V×D×P∎	P represents density (ρ).
÷ RCL M	R=V×D×P÷M∎	M represents dynamic viscosity (μ).
ENTER	R=V×D×P÷M	Ends equation.
🔁 SHOW	CK=891A LN=9	Press and hold SHOW to display checksum and length of equation. If numbers displayed match these, equation is correctly entered.

Equation 10: Reynolds Number (Handbook p. 47)

$$\mathrm{Re} = \frac{VD}{v}$$

Entered as

R=V×D÷U

Sessions PM

Disciplines General, Chemical, Environment, Mechanical

Variables The hp 33s can only recognize single Roman letters as variables. R stands for Re. Kinematic viscosity is represented by the Greek letter v (upsilon) in the Handbook, and by the Greek letter ν (nu) in PPI books. U is used here. If preferred, different letters can be assigned to the variables, but will lead to a different checksum at the end.

Keys	Display	Description
⬅ EQN RCL R ⬅ =	R=∎	Selects Equation mode and starts new equation. R represents the Reynolds number (Re).
RCL V	R=V∎	V represents velocity (v).
× RCL D	R=V×D∎	D represents diameter.
÷ RCL U	R=V×D÷U∎	U represents kinematic viscosity (v).
ENTER	R=V×D÷U	Ends equation.
⬅ SHOW	CK=E0C8 LN=7	Press and hold SHOW to display checksum and length of equation. If numbers displayed match these, equation is correctly entered.

Equation 11: Darcy-Weisbach Equation (Handbook p. 47)

$$h_f = f\frac{Lv^2}{D2g}$$

Entered as

H=F×L×SQ(V)÷D÷19.62

Sessions PM

Disciplines General, Civil, Environmental, Mechanical

Variables The hp 33s can only recognize single letters as variables. H stands for h_f.

Keys	Display	Description
⤆ EQN RCL H ⤆ =	H=∎	Selects Equation mode and starts new equation. H represents head loss (h_f).
RCL F	H=F∎	F represents friction factor (f).
× RCL L	H=F×L∎	L represents length.
× x^2	H=F×L×SQ(∎	Opens square function.
RCL V ⤆)	H=F×L×SQ(V)∎	V represents velocity (v). Closes square function.
÷ RCL D ÷	=F×L×SQ(V)÷D÷∎	D represents pipe diameter. Equation is now longer than the hp 33s display. For instructions on scrolling left and right, see How to Use This Book.
1 9 · 6 2	Q(V)÷D÷ 19.62_	19.62 is the value of $2g$. Cursor changes for entering digits.
ENTER	H=F×L×SQ(V)÷D÷	Ends equation.
⤆ SHOW	CK=3AFE LN=19	Press and hold SHOW to display checksum and length of equation. If numbers displayed match these, equation is correctly entered.

Equation 12: Venturi Meters (Handbook p. 50)

$$Q = \frac{C_v A_2}{\left(\sqrt{1 - \left(\dfrac{A_2}{A_1}\right)^2}\right)} \left(\sqrt{2g\left(\frac{p_1}{\gamma} + z_1 - \frac{p_2}{\gamma} - z_2\right)}\right)$$

Entered as

Q=C×F×SQRT(19.62×(O÷G+Y−P÷G−Z))÷SQRT(1−(SQ(F÷I)))

Sessions PM

Disciplines General, Civil, Environmental, Mechanical

Variables The hp 33s can only recognize single Roman letters as variables. To distinguish variables with subscript 1 from those with subscript 2, P and Z stand for p_2 and z_2, respectively. For p_1 and z_1, each letter is shifted back one place in the alphabet and O and Y, respectively, are used. For A_1 and A_2, the *initial* and *final* values for the area, I and F are used, respectively. G stands for the Greek letter gamma. This won't conflict with the constant $2g$, which is entered as a numerical value. If preferred, different letters can be assigned to the variables, but will lead to a different checksum at the end.

Keys	Display	Description
⬛ EQN RCL Q ⬛ =	Q=∎	Selects Equation mode and starts new equation. Q represents quantity of flow.
RCL C	Q=C∎	C represents coefficient of velocity (C_v).
× RCL F	Q=C×F∎	F stands for A_2, the final value for area, as explained above.
× √x̄	Q=C×F×SQRT(∎	Opens square root function.
1 9 · 6 2	×F×SQRT(19.62_	19.62 is the value of $2g$. Cursor changes for entering digits. Equation is now longer than the hp 33s display. For instructions on scrolling left and right, see How to Use This Book.
× ⬛ () RCL O	×SQRT(19.62×(O∎	O stands for p_1.
÷ RCL G	QRT(19.62×(O÷G∎	G represents specific weight (γ).
+ RCL Y	T(19.62×(O÷G+Y∎	Y stands for z_1.
− RCL P	19.62×(O÷G+Y−P∎	P represents pressure (p_2).

÷ RCL G	$62\times(O\div G+Y-P\div G\blacksquare$	
− RCL Z	$\times(O\div G+Y-P\div G-Z\blacksquare$	Z represents elevation (z_2).
⇄) ⇄)	$O\div G+Y-P\div G-Z))\blacksquare$	
÷ √x̄	$P\div G-Z))\div SQRT(\blacksquare$	Opens square root function.
1	$G-Z))\div SQRT(\ 1\_$	Cursor changes for entering digits.
− x²	$))\div SQRT(1-SQ(\blacksquare$	Opens square function.
RCL F ÷ RCL I	$SQRT(1-SQ(F\div I\blacksquare$	I stands for A_1, the initial value for area, as described above.
⇄) ⇄)	$RT(1-SQ(F\div I))\blacksquare$	
ENTER	$Q=C\times F\times SQRT(19.6$	Ends equation.
⇄ SHOW	$CK=1355$ $LN=47$	Press and hold SHOW to display checksum and length of equation. If numbers displayed match these, equation is correctly entered.

THERMODYNAMICS

Equation 13: Ideal Gases (Handbook p. 56)

$pv = RT$

Entered as

$P \times V = R \times T$

Sessions AM, PM

Disciplines General, Chemical, Environmental, Mechanical

Variables The hp 33s can only recognize Roman letters as variables. V represents specific volume, which is usually represented by v (upsilon) in PPI books, and by v in the NCEES Handbook. If preferred, different letters can be assigned to the variables, but will lead to a different checksum at the end.

Keys	Display	Description
🔁 EQN RCL P	P ∎	Selects Equation mode and starts new equation. *P* represents pressure (*p*).
× RCL V	P×V∎	*V* represents specific volume (*v*).
🔁 = RCL R	P×V=R∎	*R* represents the gas constant.
× RCL T	P×V=R×T∎	*T* represents temperature.
ENTER	P×V=R×T	Ends equation.
🔁 SHOW	CK=2F0A LN=7	Press and hold SHOW to display checksum and length of equation. If numbers displayed match these, equation is correctly entered.

Equation 14: Ideal Gases (Handbook p. 56)

$$pV = mRT$$

Entered as

P×V=M×R×T

Sessions AM, PM

Disciplines General, Chemical, Environment, Mechanical

Keys	Display	Description
🔁 EQN RCL P	P ∎	Selects Equation mode and starts new equation. *P* represents pressure (*p*).
× RCL V	P×V∎	*V* represents volume.
🔁 = RCL M	P×V=M∎	*M* represents mass (*m*).
× RCL R	P×V=M×R∎	*R* represents the gas constant.

Keys	Display	Description
☒ RCL T	P×V=M×R×T ■	*T* represents temperature.
ENTER	P×V=M×R×T	Ends equation.
⇄ SHOW	CK=031A LN=9	Press and hold SHOW to display checksum and length of equation. If numbers displayed match these, equation is correctly entered.

Equation 15: Ideal Gases (Handbook p. 56)

$$\frac{p_1 v_1}{T_1} = \frac{p_2 v_2}{T_2}$$

Entered as

O×U÷S=P×V÷T

Sessions AM, PM

Disciplines General, Chemical, Environment, Mechanical

Variables The hp 33s can only recognize single Roman letters as variables. V represents specific volume, which is usually represented by v (upsilon) in PPI books, and by v in the NCEES Handbook. To distinguish variables with subscript 1 from those with subscript 2, P, V, and T stand for p_2, v_2, and T_2, respectively. For p_1, v_1, and T_1, each letter is shifted back one place in the alphabet and O, U, and S, respectively, are used. If preferred, different letters can be assigned to the variables, but will lead to a different checksum at the end.

Keys	Display	Description
⇄ EQN RCL O	O ■	Selects Equation mode and starts new equation. *O* stands for p_1, as explained above.
☒ RCL U	O×U ■	*U* stands for v_1.
÷ RCL S	O×U÷S ■	*S* stands for T_1.
⇄ = RCL P	O×U÷S=P ■	*P* represents pressure (p_2).
☒ RCL V	O×U÷S=P×V ■	*V* represents specific volume (v_2).
÷ RCL T	O×U÷S=P×V÷T ■	*T* represents temperature (T_2).

[ENTER]	$0 \times U \div S = P \times V \div T$	Ends equation.
[⬅] [SHOW]	CK=9CEF LN=11	Press and hold SHOW to display checksum and length of equation. If numbers displayed match these, equation is correctly entered.

Equation 16: Constant Entropy Processes (Handbook p. 56)

$$p_1 v_1^k = p_2 v_2^k$$

Entered as

$0 \times U \char94 K = P \times V \char94 K$

Sessions AM, PM

Disciplines General, Mechanical

Variables The hp 33s can only recognize single Roman letters as variables. V represents specific volume, which is usually upsilon (v) in PPI books, and v in the NCEES Handbook. To distinguish variables with subscript 1 from those with subscript 2, P and V stand for p_2 and v_2, respectively. For p_1 and v_1, each letter is shifted back one place in the alphabet and O and U, respectively, are used. If preferred, different letters can be assigned to the variables, but will lead to a different checksum at the end.

Keys	Display	Description
[⬅] [EQN] [RCL] [O]	0■	Selects Equation mode and starts new equation. O stands for p_1, as explained above.
[×] [RCL] [U]	$0 \times U$■	U stands for v_1.
[y^x] [RCL] [K]	$0 \times U \char94 K$■	K represents the ratio of specific heats (k).
[⬅] [=] [RCL] [P]	$0 \times U \char94 K = P$■	P represents pressure (p_2).
[×] [RCL] [V]	$0 \times U \char94 K = P \times V$■	V represents specific volume (v_2).
[y^x] [RCL] [K]	$0 \times U \char94 K = P \times V \char94 K$■	
[ENTER]	$0 \times U \char94 K = P \times V \char94 K$	Ends equation.

PROFESSIONAL PUBLICATIONS, INC.

⟳ SHOW	CK=31D9		Press and hold SHOW to display checksum and length of equation. If numbers displayed match these, equation is correctly entered.
	LN=11		

Equation 17: Constant Entropy Processes (Handbook p. 56)

$$T_1 p_1^{1-k/k} = T_2 p_2^{1-k/k}$$

Entered as

S×O^((1−K)÷K)=T×P^((1−K)÷K)

Sessions AM, PM

Disciplines General, Mechanical

Variables The hp 33s can only recognize single letters as variables. To distinguish variables with subscript 1 from those with subscript 2, T and P stand for T_2 and p_2, respectively. For T_1 and p_1, each letter is shifted back one place in the alphabet and S and O, respectively, are used. If preferred, different letters can be assigned to the variables, but will lead to a different checksum at the end.

Keys	Display	Description
⟳ EQN RCL S	S∎	Selects Equation mode and starts new equation. S stands for T_1, as explained above.
× RCL O	S×O∎	O stands for p_1.
y^x ⟳ ⦅ ⟳ ⦅	S×O^((∎	
1	S×O^((1_	Cursor changes for entering digits.
− RCL K	S×O^((1−K∎	K represents the ratio of specific heats (k).
⟳ ⦆ ÷ RCL K	S×O^((1−K)÷K∎	

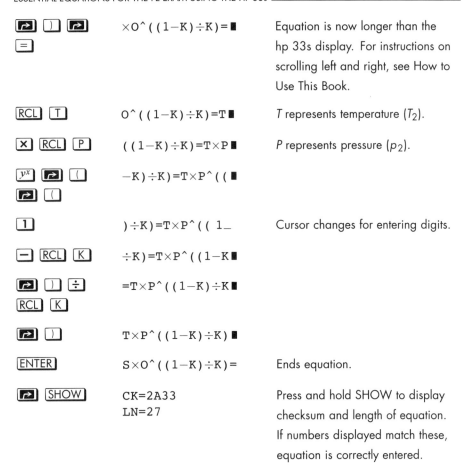

Keys	Display	Notes
🠖 ⟩ 🠖 =	$\times 0 \char94 ((1-K)\div K)=\blacksquare$	Equation is now longer than the hp 33s display. For instructions on scrolling left and right, see How to Use This Book.
RCL T	$0 \char94 ((1-K)\div K)=T\blacksquare$	T represents temperature (T_2).
× RCL P	$((1-K)\div K)=T\times P\blacksquare$	P represents pressure (p_2).
y^x 🠖 ⟨ 🠖 ⟨	$-K)\div K)=T\times P\char94((\blacksquare$	
1	$)\div K)=T\times P\char94((\ 1\_$	Cursor changes for entering digits.
− RCL K	$\div K)=T\times P\char94((1-K\blacksquare$	
🠖 ⟩ ÷ RCL K	$=T\times P\char94((1-K)\div K\blacksquare$	
🠖 ⟩	$T\times P\char94((1-K)\div K)\ \blacksquare$	
ENTER	$S\times 0\char94((1-K)\div K)=$	Ends equation.
🠖 SHOW	CK=2A33 LN=27	Press and hold SHOW to display checksum and length of equation. If numbers displayed match these, equation is correctly entered.

Equation 18: Carnot Cycle Efficiency (Handbook p. 58)

$$\eta_c = \frac{(T_H - T_L)}{T_H}$$

Entered as

$N=(H-L)\div H$

Sessions AM, PM

Disciplines General, Chemical, Electrical, Mechanical

Variables The hp 33s can only recognize single Roman letters as variables. For the Greek letter η (eta), N is used, because of the resemblance to a lowercase n. For T_H and T_L, H and L, respectively, are used. If preferred, different letters can be assigned to the variables, but will lead to a different checksum at the end.

Keys	Display	Description
⇨ EQN RCL N	N∎	Selects Equation mode and starts new equation. *N* represents efficiency (η), as explained above.
⇨ = ⇨ (	N=(∎	
RCL H	N=(H∎	*H* stands for T_H, high temperature.
− RCL L	N=(H−L∎	*L* stands for T_L, low temperature.
⇨) ÷ RCL H	N=(H−L)÷H∎	
ENTER	N=(H−L)÷H	Ends equation.
⇨ SHOW	CK=574D LN=9	Press and hold SHOW to display checksum and length of equation. If numbers displayed match these, equation is correctly entered.

HEAT TRANSFER

Equation 19: Conduction Through a Plane Wall (Handbook p. 67)

$$\dot{Q} = \frac{-kA\,(T_2 - T_1)}{L}$$

The negative sign indicates the direction of heat flow (from T_1 to T_2).

Entered as

Q=−K×A×(T−S)÷L

Sessions AM, PM

Disciplines General, Mechanical

Variables The hp 33s can only recognize single letters as variables. Q (without the dot) represents $\dot{Q}$. To distinguish subscript 1 from subscript 2, T stands for T_2, and T_1 is shifted back one place in the alphabet and S is used. If preferred, different letters can be assigned to the variables, but will lead to a different checksum at the end.

Keys	Display	Description
⟲ EQN RCL Q ⟲ =	Q=■	Selects Equation mode and starts new equation. Q stands for heat transfer rate, as explained above.
+/−	Q=−■	Changes sign of next number. Negative sign indicates direction of heat flow from T_1 to T_2.
RCL K	Q=−K■	K represents thermal conductivity (k).
× RCL A	Q=−K×A■	A represents wall surface area.
× ⟲ (RCL T	Q=−K×A×(T■	T stands for T_2, far side temperature.
− RCL S	Q=−K×A×(T−S■	S stands for T_1, near side temperature.
⟲) ÷ RCL L	=−K×A×(T−S)÷L■	L represents wall thickness. Equation is now longer than the hp 33s display. For instructions on scrolling left and right, see How to Use This Book.
ENTER	Q=−K×A×(T−S)÷L	Ends equation.
⟲ SHOW	CK=0AFB LN=14	Press and hold SHOW to display checksum and length of equation. If numbers displayed match these, equation is correctly entered.

Equation 20: Conduction Through a Cylindrical Wall (Handbook p. 67)

$$\dot{Q} = \frac{2\pi k L\,(T_1 - T_2)}{\ln\left(\dfrac{r_2}{r_1}\right)}$$

PROFESSIONAL PUBLICATIONS, INC.

Entered as

Q=2×π×K×L×(S−T)÷LN(R÷I)

Sessions PM

Disciplines General, Mechanical

Variables The hp 33s can only recognize single letters as variables. Q (without the dot) represents $\dot{Q}$. To distinguish subscript 1 from subscript 2, T stands for T_2, and T_1 is shifted back one place in the alphabet and S is used. The same cannot be done for r_1 and r_2, because Q is used elsewhere, so I (for inner radius) represents r_1, and R stands for r_2. If preferred, different letters can be assigned to the variables, but will lead to a different checksum at the end.

Keys	Display	Description
⏩ EQN RCL Q ⏩ =	Q=∎	Selects Equation mode and starts new equation. Q stands for $\dot{Q}$, as explained above.
2	Q= 2_	Cursor changes for entering digits.
× ⏩ π	Q=2×π∎	
× RCL K	Q=2×π×K∎	K represents thermal conductivity (k).
× RCL L	Q=2×π×K×L∎	L represents wall thickness.
× ⏩ (RCL S	Q=2×π×K×L×(S∎	S stands for T_1, or inner temperature.
− RCL T	=2×π×K×L×(S−T∎	T stands for T_2, or outer temperature. Equation is now longer than the hp 33s display. For instructions on scrolling left and right, see How to Use This Book.
⏩) ÷	×π×K×L×(S−T)÷∎	
LN	K×L×(S−T)÷LN(∎	Opens natural logarithm function.
RCL R	×L×(S−T)÷LN(R∎	R represents outer radius (r_2).
÷ RCL I	×(S−T)÷LN(R÷I∎	I represents inner radius (r_1).

🔁 ⟧	(S−T)÷LN(R÷I) ∎		Closes natural logarithm function.
ENTER	Q=2×π×K×L×(S−T		Ends equation.
🔁 SHOW	CK=C5B7 LN=23		Press and hold SHOW to display checksum and length of equation. If numbers displayed match these, equation is correctly entered.

Equation 21: Radiation Emitted by a Body (Handbook p. 68)

$$\dot{Q} = \varepsilon \sigma A T^4$$

Entered as

Q=E×5.67E−8×A×T^4

Sessions AM, PM

Disciplines General, Mechanical

Variables The hp 33s can only recognize Roman letters as variables. Q (without the dot) represents $\dot{Q}$. For the Greek letter ε (epsilon), E is used. The Greek letter σ (sigma) represents the Stefan-Boltzmann constant, and is replaced by its value, 5.67×10^{-8}. Exponents of 10 are introduced with the $\boxed{E}$ key; it should not be confused with the variable name E. If preferred, different letters can be assigned to the variables, but will lead to a different checksum at the end.

Keys	Display	Description
🔁 EQN RCL Q 🔁 =	Q=∎	Selects Equation mode and starts new equation. Q stands for $\dot{Q}$, as explained above.
RCL E	Q=E∎	E represents emissivity (ε).
× 5 · 6 7	Q=E× 5.67_	Cursor changes for entering digits.
E	Q=E× 5.67E_	Introduces an exponent of 10.
8 +/−	Q=E× 5.67E−8_	5.67×10^{-8} is the Stefan-Boltzmann constant (σ).
× RCL A	Q=E×5.67E−8×A∎	A represents surface area.

Keys	Display	Description
☒ RCL T	=E×5.67E−8×A×T■	*T* represents absolute temperature. Equation is now longer than the hp 33s display. For instructions on scrolling left and right, see How to Use This Book.
y^x	E×5.67E−8×A×T^■	Enters an exponent.
4	5.67E−8×A×T^ 4_	
ENTER	Q=E×5.67E−8×A×T	Ends equation.
↱ SHOW	CK=E995 LN=17	Press and hold SHOW to display checksum and length of equation. If numbers displayed match these, equation is correctly entered.

Equation 22: Black Body Radiation Energy Exchange (Handbook p. 68)

$$\dot{Q}_{12} = A_1 F_{12} \sigma \left(T_1^4 - T_2^4 \right)$$

Entered as

Q=A×F×5.67E−8×(S^4(T^4)

Sessions PM

Disciplines General, Mechanical

Variables The hp 33s can only recognize single Roman letters as variables. Q (without the dot) represents $\dot{Q}_{12}$ and subscripts are dropped from A_1 and F_{12}. The Greek letter σ (sigma) represents the Stefan-Boltzmann constant, and is replaced by its value, 5.67×10^{-8}. Exponents of 10 are introduced with the E key; it should not be confused with the variable name E. To distinguish subscript 1 from subscript 2, T stands for T_2, and T_1 is shifted back one place in the alphabet and S is used. If preferred, different letters can be assigned to the variables, but will lead to a different checksum at the end.

Keys	Display	Description
↱ EQN RCL Q ↱ =	Q=■	Selects Equation mode and starts new equation. Q stands for $\dot{Q}$, as explained above.
RCL A	Q=A■	A represents surface area (A_1).

$\boxed{\times}$ $\boxed{\text{RCL}}$ $\boxed{\text{F}}$	Q=A×F∎	*F* represents shape factor (F_{12}).
$\boxed{\times}$ $\boxed{5}$ $\boxed{\cdot}$ $\boxed{6}$ $\boxed{7}$	Q=A×F× 5.67_	Cursor changes for entering digits.
$\boxed{\text{E}}$	Q=A×F× 5.67E_	Introduces an exponent of 10.
$\boxed{8}$ $\boxed{+/-}$	Q=A×F× 5.67E-8_	5.67×10^{-8} is the Stefan-Boltzmann constant (σ).
$\boxed{\times}$ $\boxed{\rightleftarrows}$ $\boxed{(\,)}$	=A×F×5.67E-8×(∎	Equation is now longer than the hp 33s display. For instructions on scrolling left and right, see How to Use This Book.
$\boxed{\text{RCL}}$ $\boxed{\text{S}}$	A×F×5.67E-8×(S∎	*S* stands for T_1, as explained above.
$\boxed{y^x}$	×F×5.67E-8×(S^∎	Enters an exponent.
$\boxed{4}$	×5.67E-8×(S^ 4_	Cursor changes for entering digits.
$\boxed{-}$ $\boxed{\text{RCL}}$ $\boxed{\text{T}}$	5.67E-8×(S^4-T∎	*T* represents absolute temperature (T_2).
$\boxed{y^x}$ $\boxed{4}$ $\boxed{\rightleftarrows}$ $\boxed{)}$	E-8×(S^4-T^4)∎	
$\boxed{\text{ENTER}}$	Q=A×F×5.67E-8×(	Ends equation.
$\boxed{\rightleftarrows}$ $\boxed{\text{SHOW}}$	CK=66E1 LN=23	Press and hold SHOW to display checksum and length of equation. If numbers displayed match these, equation is correctly entered.

Equation 23: Body Immersed in Flowing Fluid (Handbook p. 69)

$$\mathrm{Nu} = c\mathrm{Re}^n\mathrm{Pr}^{1/3}$$

Entered as

N=C×R^O×P^0.33

Sessions PM

Disciplines Chemical, Mechanical

Variables The hp 33s can only recognize single uppercase letters as variables. N represents Nu, the Nusselt number; R represents Re, the Reynolds number; and P represents Pr, the Prandtl number. As N is now in use for something else, the next letter in the alphabet, O, is used for n. If preferred, different letters can be assigned to the variables, but will lead to a different checksum at the end.

Keys	Display	Description
▣ EQN RCL N ▣ =	N=∎	Selects Equation mode and starts new equation. N represents the Nusselt number (Nu).
RCL C	N=C∎	Enters c, a variable defined in the Handbook (p. 70).
× RCL R	N=C×R∎	R represents the Reynolds number (Re).
y^x RCL O	N=C×R^O∎	Enters an exponent. O stands for n, as explained above.
× RCL P	N=C×R^O×P∎	P represents the Prandtl number (Pr).
y^x 0 . 3 3	C×R^O×P^ 0.33_	In Equation mode, a number cannot be entered as a fraction, so 0.33 is used for 1/3. Cursor changes for entering digits. Equation is now longer than the hp 33s display. For instructions on scrolling left and right, see How to Use This Book.
ENTER	N=C×R^O×P^0.33	Ends equation.
▣ SHOW	CK=A4B5 LN=14	Press and hold SHOW to display checksum and length of equation. If numbers displayed match these, equation is correctly entered.

TRANSPORT PHENOMENA

Equation 24: Nusselt Number (Handbook p. 73)

$$\mathrm{Nu} = \frac{hD}{k}$$

Entered as

N=H×D÷K

Sessions PM

Disciplines Chemical, Mechanical

Variables The hp 33s can only recognize single Roman letters as variables. N represents Nu, the Nusselt number. If preferred, different letters can be assigned to the variables, but will lead to a different checksum at the end.

Keys	Display	Description
▶ EQN RCL N ▶ =	N=■	Selects Equation mode and starts new equation. N represents the Nusselt number (Nu).
RCL H	N=H■	H represents heat transfer coefficient (h).
× RCL D	N=H×D■	D represents inside diameter.
÷ RCL K	N=H×D÷K■	K represents thermal conductivity (k).
ENTER	N=H×D÷K	Ends equation.
▶ SHOW	CK=BDD5 LN=7	Press and hold SHOW to display checksum and length of equation. If numbers displayed match these, equation is correctly entered.

Equation 25: Prandtl Number (Handbook p. 73)

$$\mathrm{Pr} = \frac{c_P \mu}{k}$$

Entered as

P=C×M÷K

36

Sessions PM

Disciplines Chemical, Mechanical

Variables The hp 33s can only recognize single Roman letters as variables. P represents Pr, the Prandtl number. C stands for c_P, and M for the Greek letter mu (μ). If preferred, different letters can be assigned to the variables, but will lead to a different checksum at the end.

Keys	Display	Description
🔁 EQN RCL P 🔁 =	P=■	Selects Equation mode and starts new equation. P represents the Prandtl number (Pr).
RCL C	P=C■	C represents heat capacity (c_P).
× RCL M	P=C×M■	M represents viscosity (μ).
÷ RCL K	P=C×M÷K■	K represents thermal conductivity (k).
ENTER	P=C×M÷K	Ends equation.
🔁 SHOW	CK=CA48 LN=7	Press and hold SHOW to display checksum and length of equation. If numbers displayed match these, equation is correctly entered.

CHEMISTRY

Equation 26: Acids and Bases (Handbook p. 78)

$$pH = \log_{10}\left(\frac{1}{[H^+]}\right)$$

Entered as

P=LOG(1÷H)

Sessions AM, PM

Disciplines Chemical, Environmental

Variables The hp 33s can only recognize single Roman letters as variables. P represents pH, and H represents H^+. If preferred, different letters can be assigned to the variables, but will lead to a different checksum at the end.

Keys	Display	Description
☒ EQN RCL P ☒ ═	P=∎	Selects Equation mode and starts new equation. P represents pH.
◖ LOG	P=LOG(∎	Opens common logarithm function.
1	P=LOG(1_	Cursor changes for entering digits.
÷ RCL H	P=LOG(1÷H∎	H represents concentration of hydronium ions, $[H^+]$.
☒)	P=LOG(1÷H) ∎	
ENTER	P=LOG(1÷H)	Ends equation.
☒ SHOW	CK=330E LN=10	Press and hold SHOW to display checksum and length of equation. If numbers displayed match these, equation is correctly entered.

Equation 27: Faraday's Law (Handbook p. 78)

$$m_{\text{grams}} = \frac{It(\text{MW})}{(96{,}485)(\text{change in oxidation state})}$$

Entered as

M=I×T×W÷96485÷O

Sessions AM, PM

Disciplines General, Chemical, Environmental

Variables The hp 33s can only recognize single Roman letters as variables. M represents mass (m), W represents molecular weight (MW), and O represents the change in oxidation state. If preferred, different letters can be assigned to the variables, but will lead to a different checksum at the end.

Keys	Display	Description
⬛ EQN RCL M ⬛ =	M=■	Selects Equation mode and starts new equation. *M* represents mass (*m*).
RCL I	M=I■	*I* represents current.
× RCL T	M=I×T■	*T* represents time (*t*).
× RCL W	M=I×T×W■	*W* represents molecular weight (MW).
÷	M=I×T×W÷■	
9 6 4 8 5	=I×T×W÷ 96485_	Cursor changes for entering digits. Equation is now longer than the hp 33s display. For instructions on scrolling left and right, see How to Use This Book.
÷	=I×T×W÷96485÷■	
RCL O	I×T×W÷96485÷O■	*O* is change in oxidation state.
ENTER	M=I×T×W÷96485÷	Ends equation.
⬛ SHOW	CK=88C1 LN=15	Press and hold SHOW to display checksum and length of equation. If numbers displayed match these, equation is correctly entered.

MATERIALS SCIENCE/STRUCTURE OF MATTER

Equation 28: Diffusion Coefficient (Handbook p. 82)

$$D = D_0 e^{-Q/RT}$$

Entered as

D=I×EXP(−Q÷R÷T)

Sessions AM, PM

Disciplines General, Chemical, Mechanical

Variables The hp 33s can only recognize single letters as variables, so subscripts cannot be used. I represents D_0, the *initial* value of the diffusion coefficient. If preferred, different letters can be assigned to the variables, but will lead to a different checksum at the end.

Keys	Display	Description
▣ EQN RCL D ▣ =	D=∎	Selects Equation mode and starts new equation. D represents diffusion coefficient.
RCL I	D=I∎	I stands for D_0, as explained above.
× e^x	D=I×EXP(∎	Opens natural exponential function.
+/-	D=I×EXP(−∎	Changes sign of next number.
RCL Q	D=I×EXP(−Q∎	Q represents activation energy.
÷ RCL R	D=I×EXP(−Q÷R∎	R represents the gas constant.
÷ RCL T	=I×EXP(−Q÷R÷T∎	T represents absolute temperature. Equation is now longer than the hp 33s display. For instructions on scrolling left and right, see How to Use This Book.
▣)	I×EXP(−Q÷R÷T)∎	
ENTER	D=I×EXP(−Q÷R÷T	Ends equation.
▣ SHOW	CK=ED47 LN=15	Press and hold SHOW to display checksum and length of equation. If numbers displayed match these, equation is correctly entered.

Equation 29: Half-Life (Handbook p. 85)

$$N = N_0 e^{-0.0693t/\tau}$$

Entered as

N=I×EXP(−0.693×T÷H)

Sessions AM, PM

Disciplines General, Chemical, Environmental

Variables The hp 33s can only recognize single Roman letters as variables, so subscripts cannot be used. I represents N_0, the *initial* number of atoms. T cannot represent the Greek letter τ (tau), because it is used for t, time; H, for half-life, will be used instead. If preferred, different letters can be assigned to the variables, but will lead to a different checksum at the end.

Keys	Display	Description
▣ EQN RCL N ▣ =	N=∎	Selects Equation mode and starts new equation. N represents the final number of atoms.
RCL I	N=I∎	I stands for N_0, as explained above.
× e^x	N=I×EXP(∎	Opens natural exponential function.
+⁄−	N=I×EXP(−∎	Changes sign of next number.
0 · 6 9 3	I×EXP(− 0.693_	Cursor changes for entering digits. Equation is now longer than the hp 33s display. For instructions on scrolling left and right, see How to Use This Book.
× RCL T	×EXP(−0.693×T∎	T represents time (t).
÷ RCL H	XP(−0.693×T÷H∎	H represents half-life (τ).
▣)	P(−0.693×T÷H)∎	
ENTER	N=I×EXP(−0.693	Ends equation.

🔁 SHOW	CK=8D78	Press and hold SHOW to display
	LN=19	checksum and length of equation.
		If numbers displayed match these,
		equation is correctly entered.

ENGINEERING ECONOMICS

Equation 30: Single Payment Compound Amount (Handbook p. 92)

$$(F/P, i\%, n) = (1+i)^n$$

Entered as

F=(1+I)^N

Sessions AM, PM

Disciplines General, Chemical, Civil, Electrical, Industrial, Mechanical

Variables The hp 33s can only recognize single letters as variables. F represents $(F/P, i\%, n)$. If preferred, different letters can be assigned to the variables, but will lead to a different checksum at the end. Enter I as a decimal, not a percentage (for example, 0.05 instead of 5%).

Keys	Display	Description
🔁 EQN RCL F 🔁 =	F=∎	Selects Equation mode and starts new equation.
🔁 ⦅	F=(∎	
1	F=(1_	Cursor changes for entering digits.
+ RCL ⦅	F=(1+I∎	*I* represents interest rate (*i*).
🔁 ⦆ y^x	F=(1+I)^ ∎	Enters an exponent.
RCL N	F=(1+I)^N∎	*N* represents the number of periods (*n*).
ENTER	F=(1+I)^N	Ends equation.

Keys	Display	Description
🔁 SHOW	CK=67D3 LN=9	Press and hold SHOW to display checksum and length of equation. If numbers displayed match these, equation is correctly entered.

Equation 31: Capital Recovery (Handbook p. 92)

$$(A/P, i\%, n) = \frac{i\,(1+i)^{n}}{((1+i)^{n} - 1)}$$

Entered as

A=I×(1+I)^N÷((1+I)^N−1)

Sessions AM, PM

Disciplines General, Chemical, Civil, Electrical, Industrial, Mechanical

Variables The hp 33s can only recognize single letters as variables. A represents $(A/P, i\%, n)$. If preferred, different letters can be assigned to the variables, but will lead to a different checksum at the end. Enter I as a decimal, not a percentage (for example, 0.05 instead of 5%).

Keys	Display	Description
🔁 EQN RCL A 🔁 =	A=∎	Selects Equation mode and starts new equation.
RCL I	A=I∎	I represents interest rate (i).
× 🔁 (	A=I×(∎	
1	A=I×(1_	Cursor changes for entering digits.
+ RCL I 🔁)	A=I×(1+I)∎	
yˣ	A=I×(1+I)^∎	Enters an exponent.
RCL N	A=I×(1+I)^N∎	N represents number of periods (n).

Keys	Display	Description
÷ ↰ [(↰ (|	=I×(1+I)^N÷((■	Equation is now longer than the hp 33s display. For instructions on scrolling left and right, see How to Use This Book.
1 + RCL () ↰)	1+I)^N÷((1+I)■	
yˣ RCL N − 1 ↰)	N÷((1+I)^N−1)■	
ENTER	A=I×(1+I)^N÷((	Ends equation.
↰ SHOW	CK=2227 LN=23	Press and hold SHOW to display checksum and length of equation. If numbers displayed match these, equation is correctly entered.

Equation 32: Non-annual Compounding (Handbook p. 92)

$$i_e = \left(1 + \frac{r}{m}\right)^m - 1$$

Entered as

I=(1+R÷M)^M−1

Sessions AM, PM

Disciplines General, Chemical, Civil, Electrical, Industrial, Mechanical

Variables The hp 33s can only recognize single letters as variables. Therefore, the subscript in i_e is omitted. Enter R as a decimal, not a percentage (for example, 0.05 instead of 5%).

Keys	Display	Description
↰ EQN RCL () ↰ =	I=■	Selects Equation mode and starts new equation. *I* represents the effective interest rate (i_e).
↰ (|	I=(■	
1	I=(1_	Cursor changes for entering digits.

44

Keys	Display	Description
⊞ RCL R	I=(1+R■	*R* represents the nominal interest rate (*r*).
÷ RCL M	I=(1+R÷M■	*M* represents the number of periods (*m*).
⬀) y^x	I=(1+R÷M)^■	Enters an exponent.
RCL M ⊟ 1	I=(1+R÷M)^M− 1_	Cursor changes for entering digits.
ENTER	I=(1+R÷M)^M−1	Ends equation.
⬀ SHOW	CK=3CE9 LN=13	Press and hold SHOW to display checksum and length of equation. If numbers displayed match these, equation is correctly entered.

CIVIL ENGINEERING

Equation 33: Manning's Equation (Handbook p. 137)

$$Q = \frac{K}{n} AR^{2/3} S^{1/2}$$

Entered as

Q=K÷N×A×R^0.67×S^0.5

Sessions PM

Disciplines Civil

Keys	Display	Description
⬀ EQN RCL Q ⬀ =	Q=■	Selects Equation mode and starts new equation. *Q* represents discharge.
RCL K	Q=K■	*K* is a constant defined in the Handbook.

÷ RCL N	Q=K÷N∎		N represents the Manning roughness coefficient (n).
× RCL A	Q=K÷N×A∎		A represents cross-sectional area.
× RCL R	Q=K÷N×A×R∎		R represents the hydraulic radius.
y^x	Q=K÷N×A×R^∎		Enters an exponent.
0 · 6 7	K÷N×A×R^ 0.67_		In Equation mode, an exponent cannot be entered as a fraction, so 0.67 is used for 2/3. Cursor changes for entering digits. Equation is now longer than the hp 33s display. For instructions on scrolling left and right, see How to Use This Book.
× RCL S	÷N×A×R^0.67×S∎		
y^x 0 · 5	R^0.67×S^ 0.5_		0.50 is used for 1/2.
ENTER	Q=K÷N×A×R^0.67		Ends equation.
▱ SHOW	CK=D0A4 LN=20		Press and hold SHOW to display checksum and length of equation. If numbers displayed match these, equation is correctly entered.

ELECTRICAL AND COMPUTER ENGINEERING

Equation 34: Electrostatic Force (Handbook p. 167)

$$F = \frac{Q_1 Q_2}{4\pi\varepsilon r^2}$$

Entered as

F=P×Q÷4÷π÷E÷SQ(R)

Sessions AM, PM

Disciplines General, Electrical

Variables The hp 33s can only recognize single Roman letters as variables. For consistency, E represents the Greek letter ε (epsilon) in both this and the next equation. To distinguish Q_1 and Q_2, Q stands for Q_2, and Q_1 is shifted back one letter in the alphabet and P is used. If preferred, different letters can be assigned to the variables, but will lead to a different checksum at the end.

Keys	Display	Description
⟲ EQN RCL F ⟲ =	F=∎	Selects Equation mode and starts new equation. *F* represents force.
RCL P	F=P∎	*P* stands for Q_1, as explained above.
× RCL Q	F=P×Q∎	*Q* represents charge (Q_2).
÷ 4	F=P×Q÷ 4_	Cursor changes for entering digits.
÷ ⟲ π ÷ RCL E	F=P×Q÷4÷π÷E∎	*E* represents permittivity (ε).
÷ x^2	P×Q÷4÷π÷E÷SQ(∎	Opens square function. Equation is now longer than the hp 33s display. For instructions on scrolling left and right, see How to Use This Book.
RCL R ⟲)	Q÷4÷π÷E÷SQ(R) ∎	*R* represents the distance between charges. Closes square function.
ENTER	F=P×Q÷4÷π÷E÷SQ	Ends equation.
⟲ SHOW	CK=8DA6 LN=17	Press and hold SHOW to display checksum and length of equation. If numbers displayed match these, equation is correctly entered.

Equation 35: Electric Field Intensity (Handbook p. 167)

$$E = \frac{Q_1}{4\pi\varepsilon r^2}$$

Entered as

I=Q÷4÷π÷E÷SQ(R)

Sessions AM, PM

Disciplines General, Electrical

Variables The hp 33s can only recognize single Roman letters as variables. E cannot stand for both electric field intensity (E) and the Greek letter ε (epsilon). For consistency, E stands for ε in both this and the previous equation, and I stands for intensity. If preferred, different letters can be assigned to the variables, but will lead to a different checksum at the end.

Keys	Display	Description
[↱] [EQN] [RCL] [I] [↱] [=]	I=∎	Selects Equation mode and starts new equation. *I* represents electric field intensity.
[RCL] [Q]	I=Q∎	*Q* represents charge (*Q₁*).
[÷] [4]	I=Q÷ 4_	Cursor changes for entering digits.
[÷] [↱] [π] [÷] [RCL] [E]	I=Q÷4÷π÷E∎	*E* represents permittivity (*ε*).
[÷] [x²] [RCL] [R]	=Q÷4÷π÷E÷SQ(R∎	Opens square function. *R* represents distance (*r*). Equation is now longer than the hp 33s display. For instructions on scrolling left and right, see How to Use This Book.
[↱] [)]	Q÷4÷π÷E÷SQ(R)∎	Closes square function.
[ENTER]	I=Q÷4÷π÷E÷SQ(R	Ends equation.
[↱] [SHOW]	CK=9B9B LN=15	Press and hold SHOW to display checksum and length of equation. If numbers displayed match these, equation is correctly entered.

PROFESSIONAL PUBLICATIONS, INC.

Equation 36: Two Resistors in Parallel, or Two Capacitors in Series (Handbook p. 168)

$$R_P = \frac{R_1 R_2}{R_1 + R_2}$$

$$C_S = \frac{1}{\dfrac{1}{C_1} + \dfrac{1}{C_2}} = \frac{C_1 C_2}{C_1 + C_2}$$

Entered as

R=I×F÷(I+F)

Sessions AM, PM

Disciplines General, Electrical

Variables The hp 33s can only recognize single letters as variables, so subscripts cannot be used to distinguish R_P, R_1, and R_2. These variables are represented by R, I (for *initial*), and F (for *final*), respectively. If preferred, different letters can be assigned to the variables, but will lead to a different checksum at the end. The same equation can be used for two capacitors in series. In this case, R, I, and F are taken as standing for C_S, C_1, and C_2, respectively.

Keys	Display	Description
⤢ EQN RCL R ⤢ =	R=∎	Selects Equation mode and starts new equation. *R* represents total resistance (R_P).
RCL I	R=I∎	*I* stands for R_1, as explained above.
× RCL F	R=I×F∎	*F* stands for R_2.
÷ ⤢ (RCL I	R=I×F÷(I∎	
+ RCL F ⤢)	R=I×F÷(I+F)∎	
ENTER	R=I×F÷(I+F)	Ends equation.
⤢ SHOW	CK=65F4 LN=11	Press and hold SHOW to display checksum and length of equation. If numbers displayed match these, equation is correctly entered.

Equation 37: Energy Stored in an Inductor (Handbook p. 168)

$$U = \frac{Li_L^2}{2}$$

Entered as

U=L×SQ(I)÷2

Sessions AM, PM

Disciplines General, Electrical

Variables The hp 33s can only recognize single letters as variables, so the subscript in i_L is omitted.

Keys	Display	Description
⬅ EQN RCL U ⬅ =	U=∎	Selects Equation mode and starts new equation. *U* represents energy.
RCL L	U=L∎	*L* represents inductance.
× x^2	U=L×SQ(∎	Opens square function.
RCL I	U=L×SQ(I∎	*I* represents current (i_L).
⬅)	U=L×SQ(I)∎	Closes square function.
÷ 2	U=L×SQ(I)÷ 2_	Cursor changes for entering digits.
ENTER	U=L×SQ(I)÷2∎	Ends equation.
⬅ SHOW	CK=9314 LN=11	Press and hold SHOW to display checksum and length of equation. If numbers displayed match these, equation is correctly entered.

Equation 38: Resonance Frequency (Handbook p. 170)

$$\omega_0 = \frac{1}{\sqrt{LC}}$$

Entered as

W=1÷SQRT(L×C)

Sessions AM, PM

Disciplines General, Electrical

Variables The hp 33s can only recognize Roman letters as variables. W represents the Greek letter omega (ω). If preferred, different letters can be assigned to the variables, but will lead to a different checksum at the end.

Keys	Display	Description
[↱] [EQN] [RCL] [W] [↱] [=]	W=■	Selects Equation mode and starts new equation. W represents frequency (ω_0), as explained above.
[1]	W= 1_	Cursor changes for entering digits.
[÷] [√x̄]	W=1÷SQRT(■	Opens square root function.
[RCL] [L]	W=1÷SQRT(L■	L represents inductance.
[×] [RCL] [C]	W=1÷SQRT(L×C■	C represents capacitance.
[↱] [)]	W=1÷SQRT(L×C)■	Closes square root function.
[ENTER]	W=1÷SQRT(L×C)	Ends equation.
[↱] [SHOW]	CK=A4A6 LN=13	Press and hold SHOW to display checksum and length of equation. If numbers displayed match these, equation is correctly entered.

PRACTICE PROBLEMS USING STORED EQUATIONS

Part A gave instructions for storing 38 equations into the hp 33s calculator. The 20 practice problems in Part B show how to save time when solving problems by using stored equations. These problems are taken from a variety of sources (see References) and are representative of a wide spectrum of expected FE/EIT problems.

To use a stored equation, first enter the Equation mode. Scroll through the list of equations to bring the desired one into the bottom row of the display. If the equation has only one variable to the left of the equal sign and that variable is the one you want to solve for, press ENTER. To solve for any other variable, press the SOLVE key (one of the silver keys just under the display) followed by the letter name of the variable you want to solve for.

Once you've done that, the hp 33s will prompt you for the other variables. At each prompt, enter the value for that variable, followed by the $\boxed{\text{R/S}}$ key. You can include operators, so pressing $\boxed{5}$ $\boxed{\text{ENTER}}$ $\boxed{4}$ $\boxed{y^x}$ $\boxed{\text{R/S}}$ will enter the value 5^4 or 625. This can be convenient, as several of the problems in this section show.

In Part A, only the bottom line of the display is shown in the instructions (except for checksums). Here in Part B, both the top and bottom lines are shown whenever the top line is a prompt for the desired variable and the bottom line is the current value of that variable. When the current value is the one you want, you can keep it without having to enter it again.

In the displays shown in these instructions, it is assumed all variables equal zero at the start of each problem. This may not be true in actual use unless the calculator is cleared just before beginning. This does not affect the solution process.

See How to Use This Book for more information.

PROBLEM 1

Four fair coins are tossed at once. What is the probability of obtaining three heads and one tail?

    (A) 1/4
    (B) 3/8
    (C) 1/2
    (D) 3/4

SOLUTION 1

The binomial distribution equation (Equation 1, p. 7) can be used to determine the probability of three heads in four trials.

Keys	Display	Description
▶ EQN	$B=(\chi!(N) \div (\chi!(X$	Enters Equation mode. Scroll through list to bring needed equation into bottom line of display.
ENTER	N? 0.0000	Starts to solve for B. Prompts for N and displays current value.
4 R/S	X? 0.0000	Enters 4 for N. Prompts for X and displays current value.
3 R/S	P? 0.0000	Enters 3 for X. Prompts for P and displays current value.
0 · 5 R/S	Q? 0.0000	Enters 0.5 for P. Prompts for Q and displays current value.
0 · 5 R/S	B = 0.2500	Enters 0.5 for Q and solves for B.

The answer is (A).

PROBLEM 2

A force is defined by the vector $\mathbf{A} = 3.5\mathbf{i} - 1.5\mathbf{j} + 2.0\mathbf{k}$. $\mathbf{i}$, $\mathbf{j}$, and $\mathbf{k}$ are unit vectors in the x-, y-, and z-directions, respectively. What is most nearly the angle that the force makes with the positive y-axis?

(A) 20°
(B) 66°
(C) 70°
(D) 110°

SOLUTION 2

Use the formula for the resolution of a force (Equation 2, p. 10).

Keys	Display	Description
▣ EQN	R=SQRT(SQ(X)+S	Enters Equation mode. Scroll through list to bring needed equation to bottom line of display.
ENTER	X? 0.0000	Starts to solve for R. Prompts for X and displays current value.
3 · 5 R/S	Y? 0.0000	Enters 3.5 for X. Prompts for Y and displays current value.
1 · 5 +⁄₋ R/S	Z? 0.0000	Enters −1.5 for Y. Prompts for Z and displays current value.
2 R/S	R=4.3012	Enters 2 for Z and solves for R.

The magnitude of the force $\mathbf{A}$ is 4.3. The angle the force makes with the positive y-axis is

$$\theta = \cos^{-1}\left(\frac{y}{R}\right) = \cos^{-1}\left(\frac{-1.5}{4.3}\right)$$

$$= 110.4°$$

The answer is (D).

PROBLEM 3

A projectile is launched upward from level ground at an angle of 60° from the horizontal. It has an initial velocity of 45 m/s. What is most nearly the time the projectile will take to hit the ground?

  (A) 4.1 s
  (B) 5.8 s
  (C) 8.0 s
  (D) 9.5 s

SOLUTION 3

Use the formula for constant acceleration (Equation 3, p. 11). The unknown variable is T. Because T^2 occurs in the formula, two roots exist, but only one will give a sensible answer. The hp 33s will use an iterative procedure to find a root. To direct it to the right range of values, start by storing a reasonable estimate into memory location T.

Keys	Display	Description
[2] [0] [STO] [T]	20.0000	Stores estimate of 20 s in location T.

The projectile will experience acceleration only in the y-direction due to gravity. The y-component of velocity is

$$v_{0,y} = 45 \ \frac{\text{m}}{\text{s}} \sin 60° = 39 \ \text{m/s}$$

When the projectile is on the ground, s equals 0.

Keys	Display	Description
[⚫] [EQN]	S=D+V×T+A×SQ(T	Enters Equation mode. Scroll through list to bring needed equation to bottom line of display.
[SOLVE] [T]	S? 0.0000	Starts to solve for T. Prompts for S and displays current value.
[0] [R/S]	D? 0.0000	Enters 0 for S. Prompts for D and displays current value.
[0] [R/S]	V? 0.0000	Enters 0 for D. Prompts for V and displays current value.

3	9	R/S		A?	Enters 39 for *V*. Prompts for *A* and
				0.0000	displays current value.

9	·	8		T=7.9511	Enters −9.81 for *A* and solves for *T*.
1	+/−	R/S			

The answer is (C).

···

PROBLEM 4

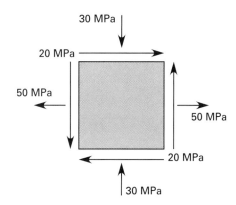

As determined from Mohr's circle, what is most nearly the maximum shear stress on the element?

- (A) 20 MPa
- (B) 42 MPa
- (C) 45 MPa
- (D) 50 MPa

SOLUTION 4

Calculate the radius of Mohr's circle (Equation 4, p. 12).

Keys	Display	Description
⬚ EQN	R=SQRT(SQ((X−Y	Enters Equation mode. Scroll through list to bring needed equation to bottom line of display.
ENTER	X? 0.0000	Starts to solve for *R*. Prompts for *X* and displays current value.

[5] [0] [R/S]	Y? 0.0000	Enters 50 for X. Prompts for Y and displays current value.
[3] [0] [+/-] [R/S]	T? 0.0000	Enters −30 for Y. Prompts for T and displays current value.
[2] [0] [R/S]	R=44.7214	Enters 20 for T and solves for R.

The answer is (C).

PROBLEM 5

What torque should be applied to the end of the steel shaft shown in order to produce a twist of 1.5°? Use 80 GPa for the shear modulus.

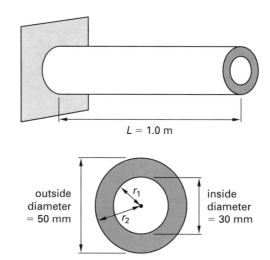

$L = 1.0$ m

outside diameter = 50 mm   inside diameter = 30 mm

(A) 420 N·M
(B) 560 N·M
(C) 830 N·M
(D) 1100 N·M

SOLUTION 5

Converting the twist angle to radians and calculating the polar moment of inertia J,

$$\phi = (1.5°)\left(\frac{2\pi \text{ rad}}{360°}\right) = 0.026 \text{ rad}$$

$$r_1 = 0.015 \text{ m}$$

$$r_2 = 0.025 \text{ m}$$

$$J = \frac{\pi}{2}\left(r_2^4 - r_1^4\right) = \left(\frac{\pi}{2}\right)\left((0.025 \text{ m})^4 - (0.015 \text{ m})^4\right)$$

$$= 5.34 \times 10^{-7} \text{ m}^4$$

Use the equation for torsional strain (Equation 5, p. 14) and solve for torque, T.

Keys	Display	Description
⮂ EQN	P=T×L÷G÷J	Enters Equation mode. Scroll through list to bring needed equation to bottom line of display.
SOLVE T	P? 0.0000	Starts to solve for T. Prompts for P and displays current value.
0 · 0 2 6 R/S	L? 0.0000	Enters 0.026 for P. Prompts for L and displays current value.
1 R/S	G? 0.0000	Enters 1 for L. Prompts for G and displays current value.
8 0 E 9 R/S	J? 0.0000	Enters 80×10^9 for Y. Prompts for J and displays current value.
5 · 3 4 E +/− 7 R/S	T=1,110.7200	Enters 5.34×10^{-7} for Y and solves for T.

The answer is (D).

PROBLEM 6

For the fixed steel rod shown, what is most nearly the force, P, necessary to deflect the rod a vertical distance of 7.5 mm?

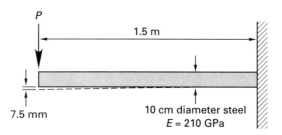

1.5 m

P

7.5 mm

10 cm diameter steel
$E = 210$ GPa

(A) 6900 N
(B) 8800 N
(C) 11 000 N
(D) 17 000 N

SOLUTION 6

Use the formula for beam deflection (Equation 6, p. 15) and solve for P. The formula for moment of inertia, I, is $\pi r^4/4$. This can be calculated during the solving process, as shown below. It could also be calculated separately before solving.

Keys	Display	Description
[⏩] [EQN]	D=P×SQ(X)×(−X+	Enters Equation mode. Scroll through list to bring needed equation to bottom line of display.
[SOLVE] [P]	D? 0.0000	Starts to solve for P. Prompts for D and displays current value.
[0] [·] [0] [0] [7] [5] [R/S]	X? 0.0000	Enters 0.0075 for D. Prompts for X and displays current value.
[1] [·] [5] [R/S]	A? 0.0000	Enters 1.5 for X. Prompts for A and displays current value.
[1] [·] [5] [R/S]	E? 0.0000	Enters 1.5 for A. Prompts for E and displays current value.

Keys	Display	Description
[2][1][0] [E]	I?	Enters 210×10^9 for E. Prompts for
[9] [R/S]	0.0000	I and displays current value.
[0] [·] [0]	I?	Enters $(0.05)^4 \times \pi \div 4$.
[5] [ENTER]	4.9087E−6	
[4] [yˣ] [↩]		
[π] [×] [4]		
[÷]		
[R/S]	P=6,872.2339	Enters value for I and solves for P.

The answer is (A).

PROBLEM 7

A pipe has a diameter of 100 mm at section AA and a diameter of 50 mm at section BB. The velocity of an incompressible fluid is 0.3 m/s at section AA. What is most nearly the flow velocity at section BB?

- (A) 0.95 m/s
- (B) 1.2 m/s
- (C) 2.1 m/s
- (D) 3.5 m/s

SOLUTION 7

Use the continuity equation (Equation 7, p. 16) and solve for V. Pressure is the same at both places; therefore, enter 1 for both O and P. The area is πr^2 in each instance. The area can be calculated during the solution process, as shown below. Alternatively, the area can be calculated prior to entering Equation mode.

Keys	Display	Description
[↩] [EQN]	P×A×V=O×Z×U	Enters Equation mode. Scroll through list to bring needed equation to bottom line of display.
[SOLVE] [V]	P? 0.0000	Starts to solve for V. Prompts for P and displays current value.
[1] [R/S]	A? 0.0000	Enters 1 for P. Prompts for A and displays current value.

⇄ π	O?	Enters $\pi \times (0.05)^2$ for A. Prompts
ENTER ·	0.0000	for O and displays current value.
0 5 x^2		
× R/S		
1 R/S	Z?	Enters 1 for O. Prompts for Z and
	0.0000	displays current value.
⇄ π	U?	Enters $\pi \times (0.10)^2$ for Z. Prompts
ENTER ·	0.0000	for U and displays current value.
1 x^2 ×		
R/S		
· 3 R/S	V=1.2000	Enters 0.3 for U and solves for V.

The answer is (B).

PROBLEM 8

Consider the holding tank shown. The tank volume remains constant. What is most nearly the velocity of the water exiting to the atmosphere?

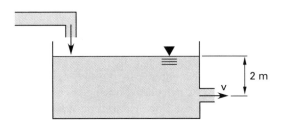

(A) 3 m/s
(B) 4 m/s
(C) 5 m/s
(D) 6 m/s

SOLUTION 8

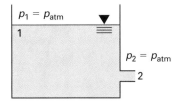

Apply Bernoulli's equation (Equation 8, p. 17) between the free surface (point 1) and the exit (point 2). Pressure is the same at both points. Therefore, p_1/ρ and p_2/ρ are equal and cancel. The most efficient way is to enter zero for both O and P. This makes the value of γ irrelevant, so 1 can be entered for G. Head loss is not a factor, so H is zero. Taking z_1 as 0 m, z_2 must be equal to -2 m. Velocity at the free surface is 0 m/s.

Keys	Display	Description
▣ EQN	P÷G+Z+SQ(V)÷19	Enters Equation mode. Scroll through list to bring needed equation to bottom line of display.
SOLVE V	P? 0.0000	Starts to solve for V. Prompts for P and displays current value.
0 R/S	G? 0.0000	Enters 0 for P. Prompts for G and displays current value.
1 R/S	Z? 0.0000	Enters 1 for G. Prompts for Z and displays current value.
2 +/− R/S	H? 0.0000	Enters −2 for Z. Prompts for H and displays current value.
0 R/S	O? 0.0000	Enters 0 for H. Prompts for O and displays current value.
0 R/S	Y? 0.0000	Enters 0 for O. Prompts for Y and displays current value.
0 R/S	U? 0.0000	Enters 0 for Y. Prompts for U and displays current value.

| ⊡ | ③ | R/S | | V=6.2714 | | Enters 0.3 for U and solves for V. |

The answer is (D).

....

PROBLEM 9

What is most nearly the head loss for water flowing through a horizontal pipe if the gage pressure at point 1 is 1.03 kPa, the gage pressure at point 2 downstream is 1.00 kPa, and the velocity is constant?

(A) 3.1×10^{-3} m
(B) 3.1×10^{-2} m
(C) 2.3×10^{-2} m
(D) 2.3 m

SOLUTION 9

Use the field equation (Equation 8, p. 17). The pipe is horizontal, so z_1 and z_2 are equal. Velocity is constant, so v_1 and v_2 are equal. The most efficient way is to enter zero for Y, Z, U, and V.

Specific weight is

$$\gamma = \rho g = \left(1000 \ \frac{\text{kg}}{\text{m}^3} \right) \left(9.81 \ \frac{\text{m}}{\text{s}^2} \right) = 9810 \ \text{kg/m}^2 \cdot \text{s}^2$$

Keys	Display	Description
▶ EQN	P÷G+Z+SQ(V)÷19	Enters Equation mode. Scroll through list to bring needed equation to bottom line of display.
SOLVE H	P? 0.0000	Starts to solve for H. Prompts for P and displays current value.
① ⓪ ⓪ ⓪ R/S	G? 0.0000	Enters 1000 for P. Prompts for G and displays current value.
⑨ ⑧ ① ⓪ R/S	Z? 0.0000	Enters 9810 for G. Prompts for Z and displays current value.
⓪ R/S	V? 0.0000	Enters 0 for Z. Prompts for V and displays current value.

[0] [R/S]	O? 0.0000	Enters 0 for V. Prompts for O and displays current value.
[1] [0] [3] [0] [R/S]	Y? 0.0000	Enters 1030 for O. Prompts for Y and displays current value.
[0] [R/S]	U? 0.0000	Enters 0 for Y. Prompts for U and displays current value.
[0] [R/S]	H=0.0031	Enters 0 for U and solves for H.

The answer is (A).

PROBLEM 10

Water is pumped from a lake to a tank along a 300 m cast iron pipeline. The inside pipe diameter is 30 cm. Minor losses, entrance losses, and exit losses are negligible. Assume steady, incompressible flow. The flow rate is 1.25 m³/s. The roughness factor for cast iron is 0.25 mm. The kinematic viscosity of water is 1×10^{-6} m²/s. What is most nearly the head loss in the pipeline?

(A) 300 m
(B) 310 m
(C) 320 m
(D) 330 m

SOLUTION 10

The relative roughness is

$$\frac{e}{D} = \frac{0.25 \text{ mm}}{(30 \text{ cm}) \left(10 \ \dfrac{\text{mm}}{\text{cm}}\right)} = 0.000833$$

The area of flow is

$$A = \frac{\pi D^2}{4} = \frac{\pi (30 \text{ cm})^2 \left(\dfrac{1 \text{ m}}{100 \text{ cm}}\right)^2}{4} = 0.07069 \text{ m}^2$$

The velocity of the flow is

$$v = \frac{Q}{A} = \frac{1.25 \ \dfrac{\text{m}^3}{\text{s}}}{0.07069 \text{ m}^2} = 17.6828 \text{ m/s}$$

Calculate the Reynolds number (Equation 10, p. 20).

Keys	Display	Description
🔲 EQN	R=V×D÷U	Enters Equation mode. Scroll through list to bring needed equation to bottom line of display.
ENTER	V? 0.0000	Starts to solve for R. Prompts for V and displays current value.
1 7 · 6 8 R/S	D? 0.0000	Enters 17.68 for V. Prompts for D and displays current value.
· 3 R/S	U? 0.0000	Enters 0.3 for D. Prompts for U and displays current value.
1 E 6 +/− R/S	R=5,304,000.000	Enters 1×10^{-6} for U and solves for R.

The Reynolds number is 5.3×10^6. From the Moody diagram, with the calculated relative roughness and the Reynolds number, the friction factor is about 0.0188. Use the Darcy-Weisbach equation (Equation 11, p. 20).

Keys	Display	Description
🔲 EQN	H=F×L×SQ(V)÷D÷	Enters Equation mode. Scroll through list to bring needed equation to bottom line of display.
ENTER	F? 0.0000	Starts to solve for H. Prompts for F and displays current value.
· 0 1 8 8 R/S	L? 0.0000	Enters 0.0188 for F. Prompts for L and displays current value.
3 0 0 R/S	V? 17.6800	Enters 300 for L. Prompts for V and displays current value.
R/S	D? 0.3000	17.68 is already value of V from previous equation. Enters 17.68 for V. Prompts for D and displays current value.

PROFESSIONAL PUBLICATIONS, INC.

Keys	Display	Description
R/S	H=299.5183	0.3 is already value of D from previous equation. Enters 0.3 for D and solves for R.

The answer is (A).

....

PROBLEM 11

An ideal gas at 0.60 atm and 87°C occupies 0.450 L. The gas constant is $R^* = 0.0821$ L·atm/mol·K. How many moles are in the sample?

(A) 0.000 20 mol
(B) 0.0091 mol
(C) 0.012 mol
(D) 0.038 mol

SOLUTION 11

Use the ideal gas law (Equation 14, p. 24). Temperature must be converted to kelvins. This can be calculated during the solution process, as shown below. It could also be calculated separately prior to entering Equation mode.

Keys	Display	Description
⏎ EQN	P×V=M×R×T	Enters Equation mode. Scroll through list to bring needed equation to bottom line of display.
SOLVE M	P? 0.0000	Starts to solve for M. Prompts for P and displays current value.
· 6 R/S	V? 0.0000	Enters 0.60 for P. Prompts for V and displays current value.
· 4 5 R/S	R? 0.0000	Enters 0.450 for V. Prompts for R and displays current value.
· 0 8 2 1 R/S	T? 0.0000	Enters 0.0821 for R. Prompts for T and displays current value.

8 7	M=0.0091	Converts 87°C to kelvins, enters
ENTER		360 for T, and solves for M.
2 7 3 +		
R/S		

The answer is (B).

....

PROBLEM 12

2 L of an ideal gas at a temperature of 25°C and a pressure of 0.101 MPa is in a 10 cm diameter cylinder with a piston at one end. The piston is depressed so the cylinder is shortened by 10 cm. The temperature increases by 2°C. What is most nearly the new pressure?

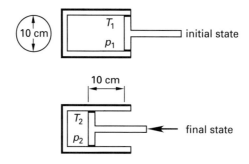

(A) 0.16 MPa
(B) 0.17 MPa
(C) 0.25 MPa
(D) 0.33 MPa

SOLUTION 12

Apply the ideal gas law (Equation 15, p. 25). Temperatures must be converted to kelvins. This can be calculated during the solution process as shown. It could also be calculated separately prior to entering Equation mode.

The original volume is 2 L, or 2000 cm³. The change in volume is

$$\Delta V = \Delta L \pi r^2 = (-10 \text{ cm})\pi(5 \text{ cm})^2 = -785 \text{ cm}^3$$

The new volume is 2000 cm³ minus 785 cm³, or 1215 cm³.

Keys	Display	Description
⟳ EQN	O×U÷S=P×V÷T	Enters Equation mode. Scroll through list to bring needed equation to bottom line of display.
SOLVE P	O? 0.0000	Starts to solve for P. Prompts for O and displays current value.
· 1 0 1 R/S	U? 0.0000	Enters 0.101 for O. Take care to use consistent units. Prompts for U and displays current value.
2 0 0 0 R/S	S? 0.0000	Enters 2000 for U. Prompts for S and displays current value.
2 5 ENTER 2 7 3 + R/S	V? 0.0000	Converts 25°C to kelvins and enters 298 for S. Prompts for V and displays current value.
1 2 1 5 R/S	T? 0.0000	Enters 1215 for V. Prompts for T and displays current value.
2 7 ENTER 2 7 3 + R/S	P=0.1674	Converts 27°C to kelvins, enters 300 for T, and solves for P.

The answer is (B).

PROBLEM 13

An 8 m long pipe of 15 cm outside diameter is covered with 2 cm of insulation with thermal conductivity of 0.09 W/m·K.

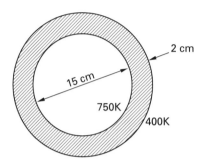

The inner and outer temperatures of the insulation are 750K and 400K, respectively. What is most nearly the rate of heat loss from the pipe?

 (A) 4.5 kW
 (B) 6.7 kW
 (C) 8.5 kW
 (D) 10 kW

SOLUTION 13

Use the formula for conduction through a cylindrical wall (Equation 20, p. 31).

Keys	Display	Description
◨ EQN	Q=2×π×K×L×(S−T	Enters Equation mode. Scroll through list to bring needed equation to bottom line of display.
ENTER	K? 0.0000	Starts to solve for Q. Prompts for K and displays current value.
· 0 9 R/S	L? 0.0000	Enters 0.09 for K. Prompts for L and displays current value.
8 R/S	S? 0.0000	Enters 8 for L. Prompts for S and displays current value.
7 5 0 R/S	T? 0.0000	Enters 750 for S. Prompts for T and displays current value.

Keys	Display	Description
[4] [0] [0] [R/S]	R? 0.0000	Enters 400 for *T*. Prompts for *R* and displays current value.
[9] [·] [5] [R/S]	I? 0.0000	Enters 9.5 for *R*. Take care to use consistent units. Prompts for *I* and displays current value.
[7] [·] [5] [R/S]	Q=6,698.1297	Enters 7.5 for *I* and solves for *Q*.

The answer is (B).

PROBLEM 14

A 3 m × 3 m plate at 500°C is suspended vertically in a very large room. The plate has an emissivity of 0.13. The room is at 25°C. What is the net heat transfer from the plate?

(A) 8.3 kW
(B) 24 kW
(C) 46 kW
(D) 47 kW

SOLUTION 14

The plate has two surfaces. Disregarding the edges, the radiating area is

$$A = (2)(3 \text{ m})(3 \text{ m}) = 18 \text{ m}^2$$

The configuration factor for a fully enclosed body is the emissivity of the body. In this case,

$$F_{12} = \epsilon_1 = 0.13$$

Use the formula for black body radiation energy exchange (Equation 22, p. 33). Temperatures must be converted to kelvins. This can be calculated during the solution process as shown. It could also be calculated separately prior to entering Equation mode.

Keys	Display	Description
[⇄] [EQN]	Q=A×F×5.67E−8×(	Enters Equation mode. Scroll through list to bring needed equation to bottom line of display.

ENTER	A?	Starts to solve for Q. Prompts for A
	0.0000	and displays current value.
1 8 R/S	F?	Enters 18 for A. Prompts for F and
	0.0000	displays current value.
. 1 3	S?	Enters 0.13 for F. Prompts for S and
R/S	0.0000	displays current value.
5 0 0	T?	Converts 500°C to kelvins and
ENTER 2	0.0000	enters 773 for S. Prompts for T and
7 3 +		displays current value.
R/S		
2 5	Q=46,325.1546	Converts 25°C to kelvins, enters
ENTER 2		298 for T, and solves for Q.
7 3 +		
R/S		

The answer is (C).

PROBLEM 15

Hydraulic fluid flows over a 2.54 cm diameter tube. The fluid flows at 13.2 m/s, normal to the longitudinal tube axis.

The following data are available for the hydraulic fluid.

$$\text{density} = 817 \text{ kg/m}^3$$
$$\text{kinematic viscosity} = 5.06 \times 10^{-6} \text{ m}^2/\text{s}$$
$$\text{heat capacity} = 2090 \text{ J/kg·k}$$
$$\text{thermal conductivity} = 0.112 \text{ W/m·K}$$

What is most nearly the convective heat transfer coefficient?

(A) 1500 kW/m²·K
(B) 3700 kW/m²·K
(C) 150 kW/m²·K
(D) 370 kW/m²·K

SOLUTION 15

The convective heat transfer coefficient can be calculated from the Nusselt number, which in turn can be calculated from the Reynolds number and the Prandtl number.

Calculate the Reynolds number (Equation 10, p. 20).

Keys	Display	Description
▣ EQN	R=V×D÷U	Enters Equation mode. Scroll through list to bring needed equation to bottom line of display.
ENTER	V? 0.0000	Starts to solve for R. Prompts for V and displays current value.
1 3 · 2 R/S	D? 0.0000	Enters 13.2 for V. Prompts for D and displays current value.
· 0 2 5 4 R/S	U? 0.0000	Enters 0.0254 for D. Take care to use consistent units. Prompts for U and displays current value.
5 · 0 6 E 6 +/− R/S	R=66,260.8696	Enters 5.06×10^{-6} for U and solves for R.

Calculate the Prandtl number (Equation 25, p. 37). Absolute dynamic viscosity (μ) is equal to kinematic viscosity times density ($\nu\rho$). This can be calculated during the solution process, as shown. It could also be calculated separately prior to entering Equation mode.

Keys	Display	Description
▣ EQN	P=C×M÷K	Enters Equation mode. Scroll through list to bring needed equation to bottom line of display.
ENTER	C? 0.0000	Starts to solve for P. Prompts for C and displays current value.
2 0 9 0 R/S	M? 0.0000	Enters 2090 for C. Prompts for M and displays current value.

Keys	Display	Description
[5] [·] [0] [6] [E] [6] [+/−] [ENTER] [8] [1] [7] [×] [R/S]	K? 0.0000	Enters 5.06×10^{-6} (kinematic viscosity) times 817 (density) for M. Prompts for K and displays current value.
[·] [1] [1] [2] [R/S]	P=77.1438	Enters 0.112 for K and solves for P.

Calculate the Nusselt number (Equation 23, p. 35). Values for C and O are found in the NCEES Handbook (p. 70).

Keys	Display	Description
[↱] [EQN]	N=C×R^O×P^0.33	Enters Equation mode. Scroll through list to bring needed equation to bottom line of display.
[ENTER]	C? 2090.0000	Starts to solve for N. Prompts for C and displays current value (from earlier equation where C meant heat capacity).
[·] [0] [2] [6] [6] [R/S]	R? 66,260.8696	Enters 0.0266 for C. Prompts for R and displays current value.
[R/S]	O? 0.0000	Retains current value for R. Prompts for O and displays current value.
[·] [8] [0] [5] [R/S]	P? 77.1438	Enters 0.805 for O. Prompts for P and displays current value.
[R/S]	N=848.7952	Retains current value for P. Solves for N.

Use the Nusselt number to calculate the convective heat transfer coefficient (Equation 24, p. 36).

Keys	Display	Description
⏩ EQN	N=H×D÷K	Enters Equation mode. Scroll through list to bring needed equation to bottom line of display.
SOLVE H	N? 848.7952	Starts to solve for H. Prompts for N and displays current value.
R/S	D? 0.0254	Retains current value for N. Prompts for D and displays current value.
R/S	K? 0.1120	Retains current value for D. Prompts for K and displays current value.
R/S	H=3,742.7189	Retains current value for K. Solves for N.

The answer is (B).

PROBLEM 16

A current of 1.5 A is supplied for 2 h to a $CuSO_4$ solution. The atomic weight of copper is 63.5. How many grams of copper will be deposited at an electrode?

- (A) 2.4 g
- (B) 3.6 g
- (C) 7.1 g
- (D) 48 g

SOLUTION 16

The dissociation reaction for copper sulfate is

$$CuSO_4 \longrightarrow Cu^{+2} + SO_4^{-2}$$

The electrolytic reaction equation is

$$Cu^{+2} + 2e \longrightarrow Cu$$

The change in oxidation number is 2 per atom of copper deposited. Use Faraday's law (Equation 27, p. 38).

Keys	Display	Description
⬛ EQN	M=I×T×W÷96485÷	Enters Equation mode. Scroll through list to bring needed equation to bottom line of display.
ENTER	I? 0.0000	Starts to solve for M. Prompts for I and displays current value.
1 · 5 R/S	T? 0.0000	Enters 1.5 for I. Prompts for T and displays current value.
2 ENTER 6 0 × 6 0 × R/S	W? 0.0000	Converts 2 h to seconds and enters 7200 for T. Prompts for W and displays current value.
6 3 · 5 R/S	O? 0.0000	Enters 63.5 for W. Prompts for O and displays current value.
2 R/S	M=3.5539	Enters 2 for O and solves for M.

The answer is (B).

PROBLEM 17

In the troposphere, ozone is produced during the day and consumed during the night. If the ozone is depleted to 10% of its initial value after 10 h of darkness, what is most nearly its half-life?

  (A) 3.0 h
  (B) 3.5 h
  (C) 4.0 h
  (D) 4.5 h

SOLUTION 17

Apply the half-life equation (Equation 29, p. 41). The final number of atoms is 10% of the initial number; choose any values such that N is 10% of I.

Keys	Display	Description
🔁 EQN	N=I×EXP(−0.693	Enters Equation mode. Scroll through list to bring needed equation to bottom line of display.
SOLVE H	N? 0.0000	Starts to solve for H. Prompts for N and displays current value.
1 R/S	I? 0.0000	Enters 1 for N. Prompts for I and displays current value.
1 0 R/S	T? 0.0000	Enters 10 for I. Prompts for T and displays current value.
1 0 R/S	H=3.0097	Enters 10 for T and solves for H.

The answer is (A).

....

PROBLEM 18

Particle A has a charge of 6×10^{-7} C, and is placed in a medium with a permittivity of 15×10^{-12} F/m. Particle B has a charge of 3×10^{-7} C, and is placed in the same medium at a distance of 1 cm from particle A. What is most nearly the force particle A exerts on particle B?

  (A) 0.33 N
  (B) 0.95 N
  (C) 3.5 N
  (D) 9.5 N

SOLUTION 18

Use Coloumb's law (Equation 34, p. 46).

Keys	Display	Description
🔁 EQN	F=P×Q÷4÷π÷E÷SQ	Enters Equation mode. Scroll through list to bring needed equation to bottom line of display.
ENTER	P? 0.0000	Starts to solve for F. Prompts for P and displays current value.

| 6 | E | 7 | | Q? | Enters 6×10^{-7} for P. Prompts for |
| +/- | R/S | | | 0.0000 | Q and displays current value. |

| 3 | E | 7 | | E? | Enters 3×10^{-7} for Q. Prompts for |
| +/- | R/S | | | 0.0000 | E and displays current value. |

1	5	E		R?	Enters 15×10^{-12} for E. Prompts
1	2	+/-		0.0000	for R and displays current value.
R/S					

| · | 0 | 1 | | F=9.5493 | Enters 0.01 for R and solves for F. |
| R/S | | | | | |

The answer is (D).

....

PROBLEM 19

What is the total resistance between points A and B?

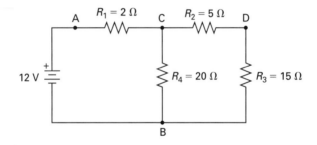

(A) $0 \ \Omega$
(B) $12 \ \Omega$
(C) $16 \ \Omega$
(D) $22 \ \Omega$

SOLUTION 19

The total resistance is the sum of the resistance between points A and C, plus the equivalent resistance of the resistors in parallel between points C and B.

$$R_{\text{total}} = R_1 + R_4 \parallel (R_2 + R_3)$$

Use the formula for two resistors in parallel (Equation 36, p. 49) to calculate $R_4 \parallel (R_2 + R_3)$. Let I be R_4, and F be $R_2 + R_3$.

78

Keys	Display	Description
🔁 EQN	R=I×F÷(I+F)	Enters Equation mode. Scroll through list to bring needed equation to bottom line of display.
ENTER	I? 0.0000	Starts to solve for R. Prompts for I and displays current value.
2 0 R/S	F? 0.0000	Enters 20 for I. Prompts for F and displays current value.
5 ENTER 1 5 + R/S	R=10.0000	Adds $R_2 + R_3$, enters the sum for F, and solves for R.

Add R_1 to this. The total resistance is 12 Ω.

The answer is (B).

PROBLEM 20

A credit card company offers students a credit line of $2000 and charges an annual percentage rate of 12%, compounded daily. What is most nearly the effective annual interest rate?

  (A) 3.28%
  (B) 12.00%
  (C) 12.75%
  (D) 13.19%

SOLUTION 20

Use the formula for non-annual compounding (Equation 32, p. 44).

Keys	Display	Description
🔁 EQN	I=(1+R÷M)^M−1	Enters Equation mode. Scroll through list to bring needed equation to bottom line of display.
ENTER	R? 0.0000	Starts to solve for I. Prompts for R and displays current value.

⌐·⌐ 1 2	M?		Enters 0.12 for R. Prompts for M
R/S	0.0000		and displays current value.

3 6 5	I=0.1275		Enters 365 for M, and solves for I.
R/S			

The answer is (C).

PRACTICE PROBLEMS USING HP 33s BUILT-IN FUNCTIONS

The instructions in Part A and the sample problems in Part B provide practice with most of the preprogrammed functions on the hp 33s one might expect to use during the FE/EIT examination. These include exponential, logarithmic, power, and trigonometric functions.

Part C demonstrates some additional functions that may be useful, involving probability and statistics, manipulating complex numbers, and converting between number systems.

The hp 33s contains other preprogrammed functions, such as the quotient and remainder, percentage, and number-altering functions, but their use on the examination is likely to be minimal.

COMPLEX NUMBERS: MULTIPLICATION

What is the product of the complex numbers $3 + 4i$ and $7 - 2i$?

- (A) $10 + 2i$
- (B) $13 + 22i$
- (C) $13 + 34i$
- (D) $29 + 22i$

SOLUTION

Keys	Display	Description
4 ENTER	4.0000 4.0000	Enters the first complex number. In RPN mode, the y-value is entered first.
3 ENTER	3.0000 3.0000	Enters the x-value.

Keys	Display	Description
[2] [+/_]	−2.0000	Enters the second complex number.
[ENTER]	−2.0000	In RPN mode, the y-value is entered first.
[7]	−2.0000 7	Enters the x-value.
[◄] [STO] [×]	22.0000 29.0000	Selects complex multiplication. The x-value of the answer appears in the X-register, and the y-value in the Y-register.

The answer is (D).

COMPLEX NUMBERS: CONVERSION FROM RECTANGULAR TO POLAR COORDINATES

What is the exponential form of the complex number $3 + 4i$?

(A) $e^{i53.1°}$
(B) $5e^{i53.1°}$
(C) $5e^{i126.9°}$
(D) $7e^{i53.1°}$

SOLUTION

The answer will be in the form $re^{i\theta}$.

Keys	Display	Description
[4] [ENTER]	4.0000 4.0000	Starts to enter the complex number. Enter the y-value first.
[3]	3.0000 3	Enters the x-value.
[◄] [→θ,r]	53.1301 5.0000	Selects conversion to polar coordinates. The r-value of the answer appears in the X-register, and the θ-value in the Y-register.

The answer is (B).

STATISTICS: MEAN AND SAMPLE STANDARD DEVIATION

What are most nearly the mean and sample standard deviation of the following numbers?

71.3, 74.0, 74.25, 78.54, 80.6

    (A) 74.3, 2.7
    (B) 74.3, 3.7
    (C) 75.0, 2.7
    (D) 75.7, 3.8

SOLUTION

Keys	Display	Description
⬅ CLEAR	1X 2VARS 3ALL 4Σ	
4	0.0000	Selects Σ to clear statistics registers.
7 1 · 3 Σ+	1.0000	Enters first of five data points.
7 4 · 0 Σ+	2.0000	
7 4 · 2 5 Σ+	3.0000	
7 8 · 5 4 Σ+	4.0000	
8 0 · 6 Σ+	5.0000	
➡ $\bar{x},\bar{y}$	$\bar{x}$ $\bar{y}$ $\bar{x}$W 75.7380	Determines mean.
➡ S,σ	sx sy σx σy 3.7557	Determines standard deviation for same data.

The answer is (D).

Selecting σx will give the population standard deviation for the data.

When two-variable data have been entered, selecting $\bar{y}$ will give the mean of the y-values, and selecting $\bar{x}$W will give the mean of the x-values using the y-values as weights. Selecting sy and σy will give the sample standard deviation and the population standard deviation, respectively, for the y-values.

PROBABILITY: COMBINATIONS

In how many ways can a 5-person committee be formed within an organization with 14 members?

    (A) 700
    (B) 1365
    (C) 2002
    (D) 5040

SOLUTION

Find the number of combinations from a set of n items, taken r at a time.

Keys	Display	Description
⬜1 ⬜4	14.0000	Enters values for n and r.
ENTER ⬜5	5	
◣ nCr	2,002.0000	Finds the number of combinations.

The answer is (C).

COMPUTER MATHEMATICS: BASE CONVERSIONS

Convert the hexadecimal number 4D3 to base-10.

    (A) 1219
    (B) 1222
    (C) 1235
    (D) 1251

SOLUTION

Keys	Display		Description
◣ BASE	1DEC	2HEX	Allows choice of base.
	3OCT	4BIN	

2	0	Selects hexadecimal. **HEX** annunciator appears.
4 D 3	4D3_	Enters 4D3.
◄ BASE	1DEC 2HEX 3OCT 4BIN	
1	1,235.0000	Selects decimal. **HEX** annunciator disappears.

The answer is (C).

....

PRACTICE PROBLEMS USING ALGEBRAIC MODE

HP users tend to use their calculators in Reverse Polish Notation or RPN mode. This mode uses less memory and tends to execute faster. Algebraic or ALG mode, however, can be easier to read and write.

When using the calculator in ALG mode, keep in mind the differences in certain functions and procedures. Most of the hp 33s User's Manual shows RPN mode, but a checkmark is given next to any step performed differently in ALG mode. Appendix C of the User's Manual also summarizes what is different in ALG mode.

This section shows a few of these differences in action, focusing on situations likely to be encountered on the FE/EIT examination.

The most common difference is in two-number operations. In RPN mode, the ENTER key is used to separate numbers. In ALG mode, this key essentially becomes the equals function—that is, the key causes the calculation to occur. For example, to raise three to the fifth power in ALG mode, one enters ⬛3⬛ ⬛y^x⬛ ⬛5⬛ ⬛ENTER⬛, whereas in RPN mode the sequence is ⬛3⬛ ⬛ENTER⬛ ⬛5⬛ ⬛y^x⬛.

In ALG mode, parentheses may be used up to 13 levels deep. The operation of the stack is different, but this is important only in programming. In RPN mode, two-variable statistical data are entered with the y-value first; in ALG mode, the x-value is entered first.

Finally, complex number arithmetic and coordinate conversions are somewhat different in the two modes. The sample problems in this section illustrate some of the differences. Each was solved in Part C using RPN mode; in this section the same problems are solved using ALG mode.

COMPLEX NUMBERS: MULTIPLICATION

What is the product of the complex numbers $3 + 4i$ and $7 - 2i$?

   (A) $10 + 2i$
   (B) $13 + 22i$
   (C) $13 + 34i$
   (D) $29 + 22i$

SOLUTION

Keys	Display	Description
[➡] [ALG]	0.0000	Shifts to Algebraic mode. **ALG** annunciator appears.
[➡] [(] [3] [+]	(3+ 3.0000	Enters real portion of first number.
[4] [◀] [CMPLX] [➡] [)]	(3+4i) RE=3.0000	Enters imaginary portion. In Algebraic mode, the [CMPLX] key is used after the imaginary part of a complex number. RE indicates the real portion; if desired, scroll down with the silver key to display the imaginary portion.
[×]	(3+4i)× RE=3.0000	This step is optional. In Algebraic mode, implied multiplication is allowed.
[➡] [(] [7] [−] [2] [◀] [CMPLX] [➡] [)]	(3+4i)×(7−2i) RE=7.0000	Enters second complex number.
[ENTER]	(3+4i)×(7−2i) RE=29.0000	Performs calculation and displays real portion of answer.
	(3+4i)×(7−2i) IM=22.0000	Scroll down with large silver key to display imaginary portion.

The answer is (D).

COMPLEX NUMBERS: CONVERSION FROM RECTANGULAR TO POLAR COORDINATES

What is the exponential form of the complex number $3 + 4i$?

(A) $e^{i53.1°}$
(B) $5e^{i53.1°}$
(C) $5e^{i126.9°}$
(D) $7e^{i53.1°}$

SOLUTION

The answer will be in the form $re^{i\theta}$.

Keys	Display	Description
[↩] [ALG]	0.0000	Shifts to Algebraic mode. **ALG** annunciator appears.
[4]	4_	Starts to enter complex number. Enter y-value first.
[x↔y]	0.0000	Exchanges X- and Y-registers.
[3]	3_	Enters x-value.
[◁] [→θ,r]	4,3→θ,r r=5.0000	Selects conversion to polar coordinates and displays r-value of answer.
	4,3→θ,r θ=53.1301	Scroll down with large silver key to display θ-value of the answer

The answer is (B).

....

PROBABILITY: COMBINATIONS

In how many ways can a 5-person committee be formed within an organization with 14 members?

(A) 700
(B) 1365
(C) 2002
(D) 5040

SOLUTION

Find the number of combinations from a set of n items, taken r at a time.

Keys	Display	Description
[↱] [ALG]	0.0000	Shifts to algebraic mode. **ALG** annunciator appears.
[1] [4] [◄] [nCr]	14nCr 14.0000	Enters value for n.
[5] [ENTER]	14nCr5= 2,002.0000	Enters value for r and finds number of combinations.

The answer is (C).

Turn to PPI for Your FE Review Materials
The Most Trusted Source for FE Exam Preparation
Visit www.ppi2pass.com today!

General FE Exam Review

FE Review Manual
Michael R. Lindeburg, PE

Engineer-In-Training Reference Manual
Michael R. Lindeburg, PE

Solutions Manual for the Engineer-In-Training Reference Manual (SI Units)
Michael R. Lindeburg, PE

FE/EIT Sample Examinations
Michael R. Lindeburg, PE

101 Solved Engineering Fundamentals Problems
Michael R. Lindeburg, PE

Discipline-Specific FE Exam Review

Chemical Discipline-Specific Review for the FE/EIT Exam
Stephanie T. Lopina, PhD, PE, with Michael R. Lindeburg, PE

Civil Discipline-Specific Review for the FE/EIT Exam
Robert H. Kim, MSCE, PE, with Michael R. Lindeburg, PE

Electrical Discipline-Specific Review for the FE/EIT Exam
Robert B. Angus, PE, et al., with Michael R. Lindeburg, PE

Environmental Discipline-Specific Review for the FE/EIT Exam
Ashok V. Naimpally, PhD, PE, and Kirsten Sinclair Rosselot, PE

Industrial Discipline-Specific Review for the FE/EIT Exam
Department of Industrial Engineering, University of Missouri–Columbia, with Michael R. Lindeburg, PE

Mechanical Discipline-Specific Review for the FE/EIT Exam
Michel Saad, PhD, PE, and Abdie H.Tabrizi, PhD, with Michael R. Lindeburg, PE